The Secret Life of Snow

GILES WHITTELL

The Secret Life

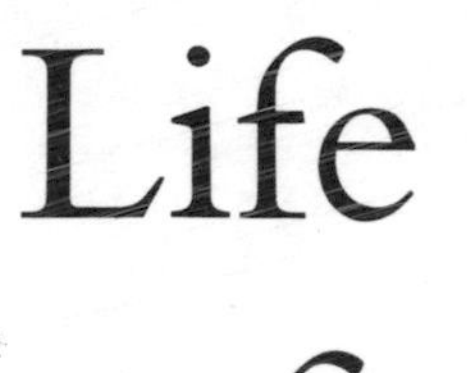

of

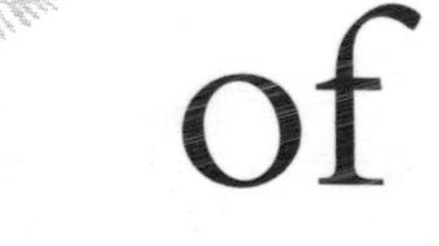

Snow

The science and the stories behind
nature's greatest wonder

First published in hardback as "*Snow: the biography*" in 2018 by Short Books
Unit 316, Screenworks, 22 Highbury Grove,
London N5 2ER

This paperback edition published in 2019

10 9 8 7 6 5 4 3 2 1

A CIP catalogue record for this book
is available from the British Library.

ISBN 978-1-78072-407-2

Cover design by Chris Bentham

Printed and bound in Great Britain
by CPI Group (UK) Ltd,Croydon, CR0 4YY

For Lucinda

who introduced me to real snow

CONTENTS

INTRODUCTION

In 1867 a sister was born to a little girl who lived on a homestead in Wisconsin. They were pioneers: home was a log cabin their father had built in woods on the north bank of the Mississippi. In the summer the woods gave shade. From November to May, like sleeping bears, they surrendered to the snow.

The family moved west as the girls grew up, but life in the snowbound woods held a special place in the younger sister's memory. Her name was Laura Ingalls Wilder and years later she described a winter's day in the cabin:

> Ma was busy all day long, cooking good things for Christmas. She baked salt-rising bread and rye'n'Injun bread, and Swedish crackers and a huge pan of baked beans, with salt pork and molasses… One morning she boiled molasses and sugar together until they made a thick syrup, and Pa brought in two pans of clean, white snow from outdoors. Laura and Mary each had a pan, and Pa and Ma showed them how to pour the dark syrup in little streams onto the snow.
> They made circles and curlicues and squiggledy things, and these hardened at once and were candy. Laura and Mary might eat one piece each, but the rest was saved for Christmas Day.

The description comes from *Little House in the Big Woods*, which Ingalls Wilder wrote in 1932 and my mother read to me in Africa. I must have been eight at the time and it made an instant, indelible impression. It was air conditioning in book form; a blast of miraculous cold in the heat of a Nigerian summer. It fixed in my mind the idea of snow as a thing of bounty.

Snow irrigates. It gives skiers something to slide on. It covers mountains from Denali to Rakaposhi like thick icing. It is the only thing on Earth that brings peace and quiet to New York City, and it makes curlicues out of molasses.

Snow has a lot in common with religion. It comes from heaven. It changes everything. It creates an alternative reality and brings on irrational behaviour in humans. There is a difference, though. Unlike religion, snow asks searching questions about the mysteries of nature.

What gives a flake its shape? Why are no two alike? How can the same warm wind bring snow to one side of a mountain and dry air to the other? How can rain sweeping up a valley past your window turn to snow in the blink of an eye?

My pleasure in moments like these is not fleeting. It can last for years, to be recalled and savoured like Proust's madeleines, and it's intensified by two things. The first is that moments of pure snow happiness are rare, especially if you live in a low and flat place like England. The second – and this, admittedly, is no more than a hunch – is that they are even more unlikely in the context of outer space.

The void that Earth hangs in is mainly a sunless, hostile vacuum. Evidence of life is scarce. Evidence of fun is even scarcer. Snow-like precipitation does seem to happen elsewhere in our galaxy, but water-based snow that you can slide down and roll around in requires a very special set of circumstances. Snow needs an atmosphere that can hold water vapour without changing its chemical composition. It needs dramatic upward movements of moist air, either over rising ground or over other, colder air masses. This movement has to lower the temperature of the moisture to freezing or below, and the air has to be naturally seeded with billions of microscopic dust particles around which ice crystals can form.

The odds against all these conditions existing in one place are high, but on Earth it happens all the time. In the thin layer of gas we call the troposphere the ingredients of snow come together routinely, as if in defiance of the cosmos. If that seems a strange thought, next time it's snowing try looking up at the clouds and picturing the void beyond them. Then go back to watching the snow as it softens hard edges, muffles all sound and turns the world a comforting, retro sort of monochrome. This simple thought experiment can put a whole new complexion on a snowstorm. What was wild and destructive becomes protective and creative. Sometimes a snowstorm can feel almost intimate. How cool is that?

Sometimes a snowstorm can be hell. It was the driving snow, not just the cold, that made Apsley Cherry-Garrard's winter trek across Ross Island in 1911 "the worst journey in the world". Everyone caught out by "bad" weather on a ski slope has experienced a version of that misery. But it's a misery to dine out on. It puts you in three-dimensional contact with nature and there is something intrinsically exciting about that. The sky is falling around you and you can taste it by sticking your tongue out. You have an excuse for being late, and for spluttering with excitement when you arrive.

Henry David Thoreau called snowflakes "glorious spangles, the sweep of Heaven's floor". He was fascinated by snow observed up close, and he was in good company. Scientists and philosophers had already been competing for three centuries to explain whether God or nature was responsible for snowflakes' six-pointed perfection. Their subject was snow in micro; snow as jewel. More recently, humans have developed an equally strong fascination with snow in macro, as commodity. There's an obvious reason for this: we cannot get enough of it. We are a thirsty species in desperate need of the water that snow stores, and we are hedonistic. When you're in snow, booted and suited to slide down it one way or another, it's natural to wonder if it can snow harder. It's the obvious question, unsubtle but insistent. How hard can snow fall?

One February morning in 1991 it snowed hard enough – in London, of all places – to save the prime minister's life. That morning a white van heading west on Whitehall pulled up on the side of the road opposite the Ministry of Defence. The driver got out and sped away on a motorbike. A few minutes later three home-made mortar shells burst through the van's fake roof. Two fell short but one landed in the garden of 10 Downing Street, 100 feet from where John Major was holding a cabinet meeting. There were a few light injuries but no one was killed – a mercy later attributed partly to the fact that the snow had hidden a mark on the pavement where the van was supposed to have stopped. The driver missed his mark.

Snow can fall hard enough to cover your tracks in the time

it takes to get inside and close the door. In 1953 the *Monthly Weather Review*, the journal of the American Meteorological Society, published a paper reassessing a storm of 1921 as the biggest one-day snow event in American history. That doesn't necessarily make it the biggest storm ever, but the New World does have what it takes to produce snow fast and in huge quantities. It has the Pacific for moisture, mountains for uplift and a continent-sized landmass for refrigeration. The storm was centred on Silver Lake, Colorado, high in the Rockies and three miles east of the continental divide. It dropped three inches an hour for 27 and a half hours straight, and 76 inches in 24 – enough to bury a six-foot-four cowboy standing up. Before that the Silver Lake event had been disqualified because of high winds and drifting, but the 1953 paper argued, in retrospect, that these were no more significant than in other mega-storms. So historic 1933 snowfalls in Maine and California were elbowed aside. Silver Lake was handed the one-day record, which it held for more than half a century.

I would give a lot to have been there, with measuring stick and fur-lined moccasins. And hip flask. And paraffin lamp flickering in the window of a log cabin stocked with enough food to last until the great spring thaw.

Will we ever see such snow again? It would be easy to look on in despair as glaciers recede above the tree line and white Christmases recede into memory, but it would be unnecessary. For many humans the experience of being in snow is so rare that it's easy to assume the same is true of snow itself. In fact, even now, we get a staggering amount of it.

Professor Kenneth Libbrecht, former head of physics at

the California Institute of Technology, has come up with a mind-bending illustration of how much snow still falls on the planet as a whole. In numerical terms he puts it at a million billion snowflakes per second, on average, every second of the year. To reach an estimate for a whole year from this starting point you would simply multiply a million billion by 60 to reach a number for each minute, then by 60 again for each hour, then by 24 for each day and finally by 365. That works out at 315,000,000,000,000,000,000,000, or 315 billion trillion snowflakes a year. Which is a big number, and impossible to visualise. So, taking 100 million snowflakes as a reasonable estimate for a modestly proportioned snowman, Libbrecht offers this thought: enough snow falls on our planet to build one snowman for every man, woman and child living on it, every ten minutes throughout the year. That's enough for seven billion snowmen every ten minutes, even in July.

Can this be true?

From any ordinary point of view it's hard to visualise enough snow falling in ten minutes to create this many snowmen, especially if you are one of the majority of humans who have never seen snow or only seen it sporadically. What's needed is an *extra*ordinary point of view, and this is what the Global Snow Lab at Rutgers University in New Jersey has found.

In 2006, Nasa launched a three-tonne satellite from Cape Canaveral aboard a Boeing Delta rocket. It's called the Geostationary Operational Environmental Satellite 13, or GOES 13, and it soars 30,000 kilometres above the North Atlantic with an unobstructed view of half the northern hemisphere. The other half is monitored and photographed by the GOES 15, parked in a similar orbit over the Pacific. Among other data streams, they provide continuously updated information on which parts of the globe are covered in snow, and the snow lab at Rutgers University turns this information into maps.

We don't know much from the Global Snow Report about the depth of this snow cover. Some of it will be thin or short-lived like the snow that came and went before Christmas in the Alps in 2016. Some of it will be deep and crisp and even, and immovable until the spring sun gets to it. What we do know is that GOES 13 mapped 50 million square kilometres of snow cover in the northern hemisphere alone in 2016–17, from the high Arctic to Anatolia. Professor Libbrecht didn't pluck his figure of a million billion flakes per second from thin air. He knew all about snow's planetary scale, and he bases his numbers on a sensible, middling estimate for the number of flakes per cubic unit of snow. Such estimates range from a few tens of millions to a billion per cubic foot, depending on the size of flake.

So he was not exaggerating about the snowmen. If all snow fell as snowmen there would indeed be a gigantic army of them, replenishing itself every few minutes. Without snow on such a scale there would be no cryosphere: no ice caps, no glaciers, nor any of the valleys they create. There would be no mountain snowpacks of the kind that store water for the summer when it's needed, from California to the Himalayas. And there would be no deep winter of the kind brought on by the reflective power of snow as it overlays the Earth like a space blanket.

Under a microscope, snow is translucent. Under the sun it's white. Much of the heat and light that hits it simply bounces back into space. This creates a feedback loop known as the albedo effect, which applies to clouds too, but more dramatically to snow: that which is cold makes the planet colder.

How much colder? Where might this feedback loop lead, other than round in circles? There is a theory, popularised in 1998 in a famous paper by the Canadian geologist Paul Hoffman, that the albedo effect of snow has in the distant past helped global glaciation reach a point of no return. About 650 million years ago, the theory goes, the annual advance of snow and ice between the poles and the lower latitudes became so exaggerated that it reached all the way to the equator and stayed there.

Welcome to snowball Earth, its surface entirely frozen over, a white planet rather than a blue one. This is a place that inspires mixed feelings for anyone who wonders as I do about the ultimate snowstorm; about why and when it happened or might happen. If snowball Earth ever existed there were probably tremendous snow events in its creation; possibly the most violent and dramatic ever. But there will have been no one around to witness them, and not much more snow once the snowball state took hold. With no open water there would have been very little evaporation or precipitation. Ice can sublimate directly to water vapour without becoming liquid, but we can be pretty sure snowball earth would not have been a good place for fresh powder. Like any normal snowball, it had two options once created: stay frozen, or melt. And since it no longer exists, if it ever did, we know it must have melted.

For a long time this was a reason to reject the whole crazy idea. If the sun's radiation would not melt the surface of the Earth once it had frozen, what would? What could reverse the feedback loop? In the early 1990s, paving the way for the Hoffman paper, a colleague of Libbrecht's by the name of Joseph Kirschvink offered an answer: volcanoes.

In a nutshell, Kirschvink suggested that massive volcanic eruptions released enough heat to melt the snowball and enough carbon dioxide to trap the heat. If he is right, this was a huge moment in the history of snow. It meant that the snowball process was repeatable. Ice could creep all the way from the poles to the equator and back, over and over again. Time lapse photography of the process would show Earth as a strange blinking eye circling the sun, gaining and losing an all-encompassing white cataract every few hundred million years.

This would solve some stubborn mysteries, such as why mineral deposits normally associated with glaciation have been found in places like Namibia. The trouble is, plate tectonics might explain these too: land masses that were once close to the poles and covered in ice have since drifted across the Earth's mantle to warmer places. So the truth about what happened 650 million years ago is that we'll never know. In practice, the oldest snowfall for which we have physical evidence is much younger. It's probably about a million and a half years old, and it sits under two miles of ice east of the Russian Vostok research station in Antarctica.

In the quest for the mightiest snowstorm of all time, Antarctica might seem an obvious place to look. It's covered in snow. It's the coldest place on Earth. It's the place where Ernest Shackleton's *Terra Nova* expedition met howling blizzards and death that came in the snow disguised as sleep. But in fact Antarctica is not snowy because a lot of snow falls there. It's snowy because so little melts. Ice cores extracted from the thickest part of the cap will reveal much about climate change over the last thousand millennia, but they're unlikely to hold evidence of the mother of all blizzards. In Antarctica it simply doesn't snow enough. The idea that it can be too cold to snow is a myth, but it can be too dry. A typical year near Vostok might see just enough new snow to cover a tennis ball.

So, for the ultimate snowfall, we have to look elsewhere. We know that snow needs moisture and something to make it freeze. For that, it turns out nothing beats a brisk wind blowing off a temperate ocean and then colliding with a range of mountains. The warmer the ocean, the more moisture it will release through evaporation, and the warmer the air above it, the more moisture it can hold. This was established in the early 19th century by a French railway engineer named Benoit Clapeyron, with the help of the German physicist Rudolf Clausius. They worked out that for every degree the sea's surface temperature rises, the atmosphere's water content should rise by 7 per cent. Several decades of measurements taken by American satellites have proved them right. Since the 1970s the world's average sea surface temperature has risen by 0.6°C, meaning the mass of atmospheric water vapour should have gone up by 4 per cent, and it has.

That 4 per cent amounts to an extra 500 cubic kilometres of water in the air around us, Dr Kevin Trenberth of the US National Center for Atmospheric Research told me a few years ago. That is one Lake Erie or three Dead Seas. Such a huge amount of water would mean more storms, he went on. "As time goes on more of these storms will be rain rather than snow, but as long as the temperatures remain low enough you may actually end up getting bigger snowstorms. We're going to have some big blizzards."

And so it has proved. For the first five years after Trenberth's prediction the Sierra Nevada mountains in California endured a dreadful drought. Then, in the winter of 2016–17, the whole range practically disappeared under snow. Squaw Valley, which claims an average of 35 feet a year, was still digging out from under 47 feet in April. That month the town announced it would stay open for skiing right through summer for the first time in its history.

Squaw had never seen anything like it. Meteorologists attributed the season's wild falls to "atmospheric rivers" supercharged with moisture thanks to the El Niño effect – a periodic warming of the Pacific that may be intensified by more general warming. If so, that warming could mean more snow before it means less, at least in a few fortunate high places. The greatest blizzard of all time may be yet to come.

But when? And where? And what sort of snow will it bring? These are serious questions for anyone who wants to be there when it happens. The answer to the first is of course unknowable, but I once asked a slightly different question – when will the *last* great blizzard happen – of the splendidly named Raymond T. Pierrehumbert, now Halley Professor of Physics at Oxford. I got a surprisingly specific answer: 2040. That was the year his statistical modelling said would yield the last big Goldilocks combination of low air temperatures, high atmospheric moisture content and heavy snow.

I hope he's wrong. We'll see. In the meantime the question of where to go for snow determines the shape of a gigantic industry – or at least you'd think it would.

From a distance, one of the most remarkable things about snow on planet Earth is the strange way the dominant local species interacts with it. Most of the 50 million square kilometres of snow in midwinter are empty of people. South of the 77th parallel (about the latitude of Thule in northern Greenland) an alien looking down from a GOE satellite starts to see humans, but only in small numbers: Inuits and oil workers on Alaska's North Slope, Saami people and snowmobiles in the north of Finland, the Inughuits of northern Greenland and the Chukchi of the Russian Far East, who on a clear day can see Alaska across the Bering Strait.

These populations are tiny and self-contained. Our alien sees little trace of them. But at the moving edge of the snow zone, from roughly the 60th parallel to the 38th, there is a frenzy of seasonal activity. Bulldozers. Building sites. High-tension cables and high-voltage power lines. Low-rise conurbations and extravagantly landscaped mountainsides. Pods, trains, funiculars and lines of steel seats suspended in mid-air, all labouring upwards against gravity so that thousands upon thousands of humans can come down with it, sliding over snow.

If snow was all that mattered for this activity, more of it would take place further north. But the frenzy happens where snow and people meet. Often it defies both nature and common sense. It speaks of a strong human compulsion.

In 2011, as the Italian economy staggered like a drunkard through the eurozone debt crisis, the Funivie Monte Bianco company committed itself to spending £100 million to upgrade a three-stage cable car that rises from the south end of the Mont Blanc tunnel to the French–Italian border. The works were wildly extravagant. They included a base station the size of a cathedral and a 150-metre pedestrian tunnel cut through solid rock to connect the lift to the Rifugio Torino, a climbers' hut two vertical kilometres above the Val d'Aosta. The cable cars themselves are circular and they rotate, so passengers don't even have to turn their heads to take in the whole alpine panorama as they ascend in a spiral towards the playground of the gods. There is a conference centre at the mid-station and a galvanised zinc summit complex like something out of *On Her Majesty's Secret Service*. The whole thing harks back half a century to the golden age of cable car construction with the unabashed aim of surpassing it at any cost.

In India 300 million people live on £1.50 or less a day but you can still ski from the highest lift in the world, ten kilometres from the highest war zone in the world. In Russia, Vladimir Putin is so beguiled by the glamour of sliding over snow that he spent $51 billion to host the winter Olympics at Sochi, a town he was brought up to love for its palm trees. In China President Xi is not to be outdone. He will host the 2022 winter games in Zhangjiakou, four hours north of Beijing near the Gobi Desert, where the only snow is artificial. Lest anyone suspect him of putting prestige before the people, Xi has ordered up 1,000 brand-new ski resorts, to be built by 2030. Most of them will be completely reliant on snow cannons.

We are mad about snow but we prefer not to travel far for it, which is a recipe for disappointment. Better to be heading for one of the small handful of places on the globe where snow still falls dry, cold, deep and often. It turns out that the quest to find them requires a little scepticism and a lot of wanderlust.

CHAPTER ONE: PERFECT SNOW

Steve: It's a snowfall. Touch it.
Wonder Woman: It's magical!
Steve: It is, isn't it?
Wonder Woman, screenplay, 2017

One January morning not long ago, the people of Ain Sefra woke to a surprise. Snow had been falling since soon after midnight on the high ground around their town. In places it was a foot deep. School was postponed so that children could go out and play, and most of them did, because Ain Sefra isn't used to snow. It's an oasis on the edge of the Sahara. Algiers is 300 miles to the north-east. The Atlantic is almost as far away to the west. Snow is rare enough here to feel like a miracle, and as the children threw themselves head first down sand dunes transformed into white-crested waves, they screeched like wild macaws.

Along the crests the snow looked as if it belonged. Lower down it quickly thinned to nothing. There was about a five-second slide before the brave sliders of Ain Sefra hit sand, at which point they ran up and did it again. With each slide they turned the snow a pinker, sandier mush. By mid-morning it was gone, but not to be forgotten. Video footage was shot and uploaded and by lunchtime was being broadcast in Brazil.

For those who felt that snow's intense coldness and its slipperiness, it must have been definitive. It was *what snow was like*. The idea that you could have different types of it would probably have seemed beside the point. Snow either fell on you or it didn't, and, praise be, it had. You could even make a case for that day's fall being the greatest snow on Earth, at least for as long as it lasted. It's the sort of case the utilitarian philosopher Jeremy Bentham might have made – the greatest possible happiness for the greatest possible number on a per-flake basis.

I like this argument, but there are a couple of problems with it. The first is that if you claim in print that Ain Sefra's snow was the greatest snow on Earth you might get sued, because "The Greatest Snow on Earth" is a trademark of the state of Utah, jealously defended since 1975. The second problem is suggested by the first: the idea of great snow isn't simple. It's contentious and the stakes are high.

The Ain Sefra fall was not a meteorological orphan. It was part of something much bigger. The main local ingredient was a long push inland by a storm system from the Atlantic. This was not unusual for the time of year, but it collided over the desert with freezing air drawn more than 3000 miles south from the Arctic, and this was a once-in-a-generation phenomenon. It was unusual for the distance travelled, the volume and temperature of the air and how long it kept coming. On its way south this air had picked up moisture from the North Sea, super-cooled the Alps and smothered them with their heaviest snow in 30 years.

If it hadn't been for this Alpine snow – which buried the glacier above Engelberg in Switzerland to a depth of five and a half metres and brought back frayed memories to a generation of old-timers who'd thought they would never see anything like it again – it's possible that Ain Sefra's sideshow would not have attracted much attention. As it was, the arrival of snow in the Sahara became supporting evidence for excited theories of a big shift in weather patterns.

The French national weather service announced the *Retour de l'Est*, an exotic name for easterly winds from Siberia by way of the Black Sea and the Med that were colliding with the Alps from the east as the Altantic system arrived from the north west. Anglo-Saxon weather watchers were preoccupied with a high pressure zone over Greenland that was forcing the jetstream into a northern detour before its arrival in Europe. And then there was the North Atlantic Oscillation, the NAO, an atmospheric see-saw powered by the Azores high and low pressure centred on Iceland. When both are weak the NAO is negative and it brings fewer storms to Europe from the Atlantic. When they are strong it's positive and that means heavy weather.

In January 2018 the NAO was strongly positive. Ordinarily the precipitation expected as a result would be relatively warm, but the *Retour de l'Est* and the Greenland high were cooling it down. Hence the excitement, and the snow.

It came a metre at a time: mighty dumps, one after the other. Cold fronts queued up over the North Atlantic and then rolled over the British Isles and the Low Countries, right up to the watershed once crossed by Hannibal. And there they sat, emptying, like supertankers caught on a reef.

I printed 20 pages of reports from one of my favourite snow sites and just sat there looking at them. Pictures sent in by tourist offices and European weather services were all variations on the theme of burial. Buried trucks. Buried chairlifts. Buried buildings. All surfaces were merely pedestals for natural art installations. All efforts to preserve the rhythms and routines of life were vanquished.

"Crazy snow depths in Cervinia."

"Maximum avalanche alert."

"Exceptionally heavy and potentially disruptive snow…"

"Chaos in the Alps!"

Captions to make the heart sing. Zermatt was cut off. Visitors to Tignes were banned from walking outside in case snow falling from roofs should bury them alive. Donald Trump's fleet of helicopters arrived in Davos to headlines of "Apocalypse Snow". Ho-hum memories of the year before were banished, and for once Europeans could pretend it snowed in their mountains as it did in America's. Which was a romantic thought, but not quite true.

A once great snow power, Europe is now managing long-term decline. Meanwhile the basic geography of the western

United States, with westerly prevailing winds off the world's largest ocean, makes it hard to beat as a snow factory. 2017 was an exception to this rule and even then Europe was playing catch-up: the previous year, the combination of a warming Pacific and a stubborn knot of mountains on the California–Nevada state line had produced something to put the Alps in their place.

Winter came late to the Sierra Nevada in 2016. There hadn't been much snow for Christmas, but by early January a tell-tale pattern was emerging in satellite measurements of water vapour in the air above the eastern Pacific. Vapour was dense in a band 100 miles wide, starting at a point roughly equidistant from Hawaii and Acapulco. With due respect to the ocean, this was in the middle of nowhere. But it was the place where two giant low-pressure zones rotating in opposite directions were drawing moisture into a long conveyor like dough into a pasta maker. And from here the band of moisture stretched 2,000 miles north-east to the Californian coast. It made landfall south of San Francisco and spread out as it hit rising ground that would eventually freeze it and turn it into snow.

This weather pattern used to be called a pineapple express. It comes from the tropics and even shows up as a green and yellow line on some of the National Oceanic and Atmospheric Administration (NOAA)'s graphics. Nowadays it's more fashionably known as an atmospheric river, although the term derives from research done in the 1990s by two scientists, Yong Zhu and Reginald Newell, who preferred to talk about tropospheric rivers.

Zhu and Newell's choice of words was revealing. The troposphere is the lowest layer of the atmosphere. The lower a body of air is, the warmer it tends to be, the more moisture it can hold, and the more likely it is to collide sooner or later with mountains.

Almost all the water vapour being fed into the great atmospheric river of January 2017 was carried along at or below 6,000 feet above sea level. Some of it condensed out as rain as it crossed the coast, watering the gardens of Carmel and the citrus groves of the San Joaquin Valley. The rest of it was going to have to deal with the Sierras.

This is how California and the Pacific collaborate to make snowstorms. Many variables have to work in synch for a big storm. When they don't, the snowpack thins and the people of Los Angeles have to sacrifice their lawns, but when atmospheric rivers flow in earnest, they can carry 15 times as much water as the Mississippi, according to the NOAA.

"They come at you like a firehose," a meteorologist from NASA told the Associated Press as the snow started to fall.

In six days, between January 6th and January 11th, enough precipitation landed on the Californian Sierras to refill more than 80 reservoirs with a combined volume of more than 40 cubic kilometres of water by the time it had all melted. In those six days Mammoth Mountain, which sits in a notch in the High Sierras that funnels weather to it from north and south, got 15 feet of snow. For the rest of January and much of February the firehose kept spraying. Squaw Valley would grab the headlines at the end of the season by announcing plans to keep its ski lifts open all summer, but it was a smaller resort on the other side of the lake that was buried deepest. Mount Rose, a steep drive up into the mountains from Reno, had measured a total of 64 feet or nearly 20 metres of snow by the time it closed its lifts.

Twenty metres of snow is almost too much to fathom. It's three two-storey houses on top of each other. I phoned the Mount Rose resort office to ask what it had been like and Mike Pierce, head of marketing, said: "Atmospheric rivers became kind of the norm. Ba-boom, ba-boom, ba-boom, ba-boom. There are times when if you didn't want it so much you'd say it was too deep. It felt bottomless."

I have good memories of atmospheric rivers even though they weren't called that at the time. In 1996 I sat for four days watching one dump snow on Mammoth, summoning the courage to ask a fellow snow addict to marry me. She said yes, so maybe that snow was perfect in its way. Maybe. But if you fall too hard for fallen snow the risk is that you stop paying attention to the snowflakes.

In the film that bears her name, Wonder Woman encounters her first snowfall in a town square half destroyed by war. She's transfixed by it. She calls it magical, and she's right. Snow is magical, not least in the sense of being mysterious. Incomprehension is justified when you see it fall. Several billion years have passed since the first flake. Half a century ago we went to the moon. We can edit genes and create membranes a single atom thick, but we still don't know how snowflakes grow.

Two hundred miles south of Mammoth Mountain a separate range rises over the Mojave Desert, and on its north-facing slopes, for a few dozen days a year, it snows. Most of what falls settles among well-spaced Douglas firs at about 8,000 feet, cooling the Earth and putting smiles on the faces of those who know where to find it.

For several years this was the closest snow to where we lived. We'd drive up out of the heat of LA and marvel at it, unaware that on the way we'd passed someone doing more than marvelling.

The road to these mountains passes through Pasadena, California, home of Caltech, where Ken Libbrecht spends much of his time listening to deep space for evidence of neutron stars and gravitational waves. His passion, though, is snowflakes. Most of his original research is on the mysteries of snow crystal morphology. At Caltech, using machines he's built himself, he has grown the world's most perfect artificial snowflakes and come up with the fullest account yet of how they grow.

Ken's shopping list for snow would look something like this:

- Water
- Air
- Refridgeration
- Temperature control
- Optics
- Dust

The formation of a snowflake starts with a few billion water molecules. In any normal cloud these cluster naturally in tiny droplets, which aren't vapour but are small enough to defy gravity. They just float there, held aloft by air. The molecules in them jostle together like shoppers on Oxford Street, close to one another but not in orderly rows. And – it's important to note – they have a natural fondness for this liquid state.

When clouds rise over banks of denser air or up the side of a mountain, they cool. In theory the droplets in them should freeze solid when the air temperature reaches 0°C, but strangely this doesn't happen. Without something to freeze onto, pure water can stay in a super-cooled liquid state down to minus 40°C. Even *with* something to latch onto a droplet in a cloud might not freeze until the air temperature falls to about minus 6°C.*

This is where dust comes in handy. Tiny specs of it are perfect nuclei for ice crystals, and there are plenty of them in the atmosphere. Volcanoes are good sources; so are forest fires, winds as they scour the Earth, and slaughterhouses. In January 2011 an unusual snowfall near Dodge City, Kansas, was attributed by the US National Weather Service to steam and soot rising from a pair of abattoirs south-east of the city. A nearby power station provided extra steam, and all three buildings lined up neatly with the prevailing south-easterly wind. Half an inch of snow fell in an expanding plume north-east of the city, exactly where

---

* Not that weather forecasters rely on temperature to predict snow. They also study pressure as a proxy for temperature. Specifically, they measure the vertical distance between the point in the atmosphere where the pressure is 1,000 millibars and the point, always some distance above, where it has fallen to 500. The colder the air, the shorter that distance, because cold air takes up less space than warm. When that distance falls to 5,400 metres or less it indicates a likelihood of snow. Lop off a zero for brevity, and you can see why snow pros talk about the "540 line" as the single most important thing to look for on a weather map.

you would expect in the circumstances, and nowhere else. The same thing happened downwind of a Pennsylvania nuclear power plant two years later, and downwind of a sewage works two years after that. In the same way, more or less, Siberians say they can make snow in the winter just by emptying a pot of lukewarm water out of the kitchen window. Presumably it helps if they live a few floors up.

Usually snowflake formation is more poetic than this, but it always starts with an ice crystal, and the crystal always has six sides.

Why six? Before the triumph of science this sort of question led into the treacherous territory between nature and the divine. Undaunted, in 1611 the German daydreamer Johannes Kepler published an entire pamphlet on it. *Cum perpetuum hoc sit*, he mused, *quoties ningere incipit, ut prima illa nivis elementa figuram prae se ferant asterisci sexanguli, causam certam esse necesse est*. "There must be some reason why, whenever snow begins to fall, its initial formations invariably display the shape of a six-cornered starlet."

Strictly speaking his question was about ice crystals rather than snow crystals, but it amounted to the same thing and he failed to answer it. He pondered nature's talent for what is now called close packing, in pea-pods, pomegranates and honeycombs. He wondered if snowflakes incorporated the same logic as cannonball stackers who arranged their balls in pyramids. In a sense they do, but he was going on little more than a hunch.

Four centuries on we can do better. The six-sidedness of ice crystals is a result of the natural angle formed by two hydrogen atoms and one oxygen atom in a molecule of water. That angle is 108 degrees. It doesn't change when water freezes but it does force water molecules into a three-dimensional hexagonal lattice as they lose energy and surrender to the cold.

A team from the Max Planck Institute in Munich has worked out that the minimum number of water molecules needed to form a stable frozen lattice is 275. The minimum number for the hexagonal shape of an ice crystal to start to emerge is more like 1,000, which is still far too small to see, but large enough for more molecules to stick to it and make it grow. As it does, it graduates from ice crystal to snow crystal, and eventually to snowflake.

The whole process is wonderfully unrushed. A snowflake can take three-quarters of an hour to fall to the ground, growing all the time and sucking vapour out of the air as it does. Ken Libbrecht explained how in a paper in 2007:

> The liquid droplets in the cloud that remain unfrozen slowly evaporate, supplying the air with the water vapour that creates their frozen brethren. Thus there is a net transfer of water molecules from liquid water to water vapour to snow crystals. This is the round-about method by which the liquid water in a cloud freezes.

Libbrecht estimates that it takes a million droplets to provide the water vapour for a single snow crystal the size of the full stop at the end of this sentence. At this stage our proto-flake is compact, streamlined and hard to distinguish from its neighbours. They would even be hard to tell apart under an electron microscope because snow's basic hexagonal lattice is uniform.

If this sounds close to saying all snow is alike, it should. At the molecular level snow is predictably structured, although only up to a point. It is also true – indisputably, amazingly – that no two flakes are alike.

Even at this full stop-sized stage, there are already two reasons why it's impossible for any two snow crystals, never mind snowflakes, to be identical. The first is that speck of dust. No two specks are created equal.

The second is deuterium, or, more precisely, deuterium hydroxide. Anyone who's seen Kirk Douglas as Dr Rolf Pedersen in the 1965 film *Heroes of Telemark* will know that the Nazis hoped to vanquish all resistance and establish their thousand-year Reich with the help of deuterium hydroxide, a naturally occurring heavy isotope of water which if isolated can be helpful in the building of an atom bomb. Left to itself, however, in water, water vapour or ice, deuterium hydroxide is dilute and harmless, and its distribution is random. There is no pattern or predictability to the whereabouts of its molecules. All we know is that there is one per 3,200 normal water molecules, which equates to several million per flake – in, remember, a totally random formation. The number of different possible arrangements of these deuterium molecules in any given flake far exceeds the number of atoms in the universe. This is the second engine of uniqueness in snowflake formation.

And the third is the atmosphere. As the crystal falls it grows. How it grows depends on what it falls through – the water content of the cloud, the temperature and pressure of the air, the route it takes. All of this is random too and randomness breeds yet more uniqueness. No two snowflakes since the dawn of snow can possibly have formed from identical ice crystals and taken exactly the same path to Earth.

This path determines a flake's shape, so the shape is a travelogue; a record of turns taken, to be read outwards from the centre. One day a supercomputer should be able to look at a flake and describe in detail every layer of air it fell through.

Even to imagine this is progress: a century ago, when a New Hampshire farmer's boy named Wilson Bentley pioneered the art of snowflake microphotography he thought snowflakes' uniqueness enhanced their beauty, but he couldn't account for it. He just observed it, over and over again, from under a hood, peering into a large-format box camera pointed at the sky as a source of light. His pictures have never been improved on, but they are less mysterious now than when he took them.

We now know that snow crystals grow into snowflakes because of three processes acting on them at once. The first two are known as faceting and branching. Imagine the ice crystal building block of any flake as a miniature hexagonal ice hockey puck. Its six sides are prism facets. Its top and bottom are its basal facets. As it falls through cold, moist air, or swirls around in it heedless of gravity, free-floating water molecules from evaporated droplets adhere directly to these facets without passing back through the liquid phase. When they stick to the basal facets the puck thickens. It can quickly become taller than it is wide – a column rather than a puck. When they stick to its prism facets it becomes wider, like a plate, and one of the mysteries of faceting is that both versions can act on one flake but they very rarely do so at the same time. This is there are snowflakes that look like pairs of train wheels on an axle. They grow first through basal faceting to create a column, then through prism faceting to create the wheels, with an abrupt switch from one to the other than no one can quite explain.

For all the oddness of train-wheel shapes, snowflakes generally would be dull if they were shaped only by faceting. They would be small and granular and lie heavy on the ground, like the rough little crystals that pile up next to snow cannons on denuded ski slopes.

What makes snowflakes beautiful is branching. In a crystal's journey through the atmosphere all the parts of its surface are in competition with each other for free-floating water molecules.* In this "competition", to stick out into the air is to have the edge, so when the prism facets are growing, the corners between them grow faster than the facets themselves. They are the equivalent of the Tweeter who gets more followers simply by posting more tweets. As the corners begin to stick out they catch even more molecules relative to the self-effacing facets, and so they grow faster – a positive feedback loop that creates the branches that turn crystals into stars and snow into something light, fluffy and miraculous.

Not all snow has such a happy destiny. I've seen it rise up in sheets of slush from the streets of Moscow, soaking whole bus stops full of bystanders. I've read about it smothering Buffalo,

* This is not strictly true. Snowflakes aren't alive. They have no agency. There is no competition in the human or animal sense either between flakes or different parts of individual flakes, but the strange thing is they strive and grow and outdo each other as if there were. The effect is the same, which is why scientists sometimes use these turns of phrase when trying to explain what's going on to the general reader, and why I have too.

New York, like an ice bucket challenge, as wet and heavy as Lake Erie itself and useless for anything except being hauled away. That said, it's surely wrong to discriminate against snowflakes on the basis of where they fall. It's not their fault.

Given the chance, all snowflakes will branch, and branching is only the beginning. In fact if branching were all that happened during pauses in faceting, snowflakes would be tangles of icy white filaments almost infinitely long. It turns out, however, that the process is self-limiting and decorative. The feedback loop that creates branches from the corners of a hexagon slows down almost as soon as it speeds up. Growth happens in fits and starts because of the microphysics of diffusion: the more successful a branch is at snaring free-floating water molecules from the air, the greater the distance to the branch from others in the immediate neighbourhood, and the longer they take to cover that distance.

This time a real-world analogy works better than Twitter. A garrulous guest at a party attracts a circle of curious listeners, creating space between them and the rest of the people in the room – space that they don't bother to cross because they can't hear what the fuss is about. However entertaining the guest, the seeds of the limits of her popularity lie in her popularity. In the jargon, her popularity is diffusion-limited. That could change, but only if she stands on a chair and starts to sing, or performs a twirl in a costume that changes as if by magic, causing a stampede – and snowflakes, of course, can do this too. Their growth produces fabulous, attention-getting patterns because the process switches back and forth between branching and faceting as they fall through layers of air of differing humidity.

Experiments with artificial snowflakes have shown that high humidity favours branching and low humidity favours faceting. Molecules adhere accordingly, like resin granules in a 3D printer.

No party is complete without a log fire or a heat lamp, or a swimming pool. And so it is with snowflakes. The final, crucial factor that determines how big and beautiful they grow is temperature. Is there a sweet spot in terms of temperature and average humidity that tends to produce the greatest snow on Earth? It turns out there is.

In 1928, after a long journey by steamship via Singapore and Suez, a young post-doctoral researcher in experimental physics arrived in London to immerse himself in the study of a revolutionary new method for photographing bones through human flesh. His name was Ukichiro Nakaya, and he was determined to succeed. Ambition, more than a specific interest in X-rays, brought him to London. Ambition took him back to his native Japan two years later to take up an assistant professorship at Hokkaido University, where at first he worried about how to make his mark. There was no money or equipment with which to compete in fashionable fields like quantum mechanics or cosmology. There was, however, super-abundant snow.

It is impossible not to envy Japan's snow geography. Sapporo sits across the sea from Vladivostok. The wind arrives chilled by 3,000 miles of Siberia, then moistened and warmed, but not too much, by 200 miles of water. Nakaya set to work photographing the results.

Aware of Bentley's work, he picked up where it left off. He took 3000 snowflake photomicrographs, measured and classified them and then started making his own snow in a temperature-controlled cloud chamber. He found the tip of a single strand of rabbit fur to be the ideal nucleation point, and grew flakes in varying humidities and temperatures down to minus 35°C . Radically different snowflake shapes formed as he played with these two variables. He grouped them into 14 categories and placed them on a graph with humidity increasing up the x axis and temperature falling along the y. The diagram this produced looks random. It is counter-intuitive. It shows no easy trend towards bigness or shapeliness in any direction, but it has remained largely unchanged for 85 years. It bears his name: the Nakaya Diagram, and it shows as clearly as a traffic light what the conditions are for perfect snow. You need a supersaturation level of 0.3 grams of water per cubic metre of air, and a temperature of minus 15°C.

This gives the best odds of snowflakes that look the part; big and beautiful but not likely to collapse under their own weight or melt on impact. These are fern-like stellar dendrites, the stars of snow.

The Nakaya Diagram shows the conditions required for columns, needles, plates and solid prism flakes as well, and I would take any of them over rain or nothing, but stellar dendrites are special. They are like ordinary dendrites, which occur at 0 to minus 3°C, but they're up to three times bigger. These are the flakes that Bentley liked best for close-ups. They settle fast, pile high and absorb more sound than any other type of snow. If you were to lie down in a snowstorm they are the type of flake you'd want to see falling towards your face. If rain is a rushed and irritable waiter nearing the end of his shift, an overnight fall of fern-like stellar dendrites is a stealth team of *metteurs-en-scene* who spread out through the banquet hall with dishes under silver domes which they remove with an understated flourish, whispering as one, "*voilà!*".

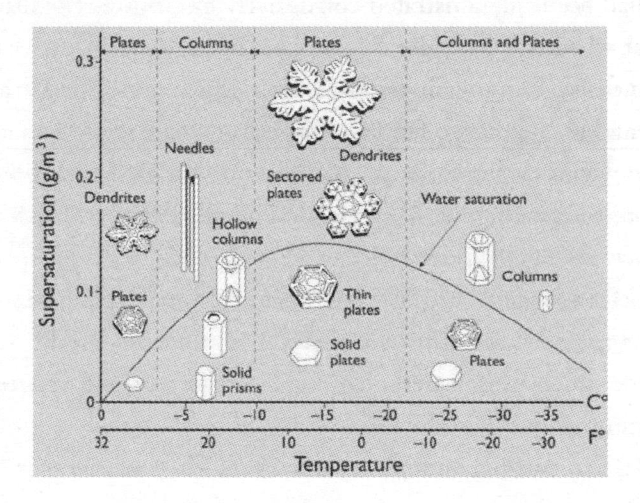

And we still don't know exactly how they grow. The Nakaya Diagram is a triumph of observation over hunch. No one could have made it up. It is the "what" of snowflake formation. Even now, we're not entirely sure about the "why". The mystery is why you might find a stellar dendrite at minus 12°C but not at, say, minus 8. Broadly speaking, the relationship between humidity and shape is understood. The more moisture there is available, the more elaborate and star-like flakes can become. But the relationship between temperature and shape is trickier. The diagram shows rapid switches from flakes that are basically column-shaped to flakes that are basically star-shaped, and back, with differences of only one or two degrees.

A quarter of a century after Nakaya's works a British scientist claimed to have understood it. In an afterword to a 1966 translation of Kepler's "The Six-Cornered Snowflake", Professor EB Mason of Imperial College London, wrote, slightly smugly, that it had been "demonstrated conclusively by the present author" that whether a flake grew as a plate, prism or star was "determined basically by air temperature". This was true as far as it went, but it didn't go far. For temperature to be the main factor you would expect to see a big difference between growth rates by prism faceting (i.e. outward) and basal faceting (i.e. upward) at a given temperature.

Mason had taken some rudimentary measurements in the 1960s, and been satisfied. Fifty years later Libbrecht took some much more careful ones and found there was no temperature-dependent difference between basal and prism growth rates that was big enough to explain the sheer size of fern-like stellar dendrites. Something else had to be at work, especially considering the effects of diffusion-limited branching, which stops branches growing too far.

Libbrecht came up with a theory. He suggested that once prism faceting had started the process of making a stellar flake, in the right conditions "the extremely narrow facet surface" at the edge of the flake "will grow much faster than a broader facet". It was all about the shape, in other words; in this case a shape characterised by extreme thinness that sets up another feedback loop producing branches that are also very thin and intricate. He called this phenomenon the *knife edge instability*, which has a certain ring to it. Then, in almost the same breath, he admitted that "the molecular dynamics of how this happens are still unclear".

That was in 2007. Ten years later I phoned him to ask if he'd managed to firm up the knife edge instability theory. The short answer was no. Snowflake science was "forbidden science", he said. Still no budget. Still no glamour. But he was still sticking to the theory. "I think the instability exists, but why? Who knows?"

And where? Libbrecht would find few fern-like stellar dendrites up behind his lab on Mount Waterman, or in Ain Sefra, come to that. They aren't cold enough. Usually the same is true in the Californian Sierras, although Mike Pierce at Mount Rose says sometimes the wind backs to the east and the mercury drops by 10°C and brings "cold smoke", too light to throw or even shovel. Libbrecht's favourite snow town is Cochrane, in northern Ontario, where the town mascot is a polar bear and the average temperature from December to March is minus 14.5°C. Winds are light and the terrain is flat; flake-catching country for the purist.

When taking photographs for science, quality is obviously the most important factor. But snow addicts crave quantity as well. All mountains can force water vapour to condense, thanks to the cooling power of orographic uplift, but some do it with special panache. Steamboat Springs, tucked in the lee of the front range of the Rockies, sought a federal trademark for the words "champagne powder" in 2008, claiming they were first used there by a local named Joe McElroy, who said his nose tickled as if from bubbles when skiing through it. True or not, the phrase captures the ideas of depth and perfection pretty well. Letters from Steamboat's lawyers defending the trademark have since been sent to most rival American resorts and have earned some intemperate replies ("Are you frickin' kidding me?" the *Aspen Times* asked in January 2010), but Libbrecht cautions against dismissing the idea that Steamboat's snow may in fact be special. There is no accounting for microclimates, especially in the mountains.

Across the tepid Atlantic the twin Austrian villages of Warth and Schröcken hold the European average seasonal snowfall record of 10.5 metres, even though they aren't widely known for it or at very high altitude. The same rule on microclimates applies,: they sit in a funnel in northern Vorarlberg in direct line of sight for every nor'wester rolling in across Lake Constance. But the average temperature and water content of Warth's snow, like so much of what falls in a warming world, would place it on the left-hand side of the Nakaya Diagram; the soggy side. This is plentiful snow, but it is far from perfect.

I always hoped to find that Kyrgyzstan, the pearl of Central Asia, would be blessed with winter snow to rival Utah's, since it occupies an equivalent place smack in the middle of its continent. But no such luck. A group of ski-tourers reporting back for the 2017 edition of the *Alpine Journal* wrote that "the lack of snow is ever more pronounced in the [Kyrgyz] Tien Shan".

Which brings us to Utah itself, as destiny brought the Mormons. "This is the place," said Brigham Young of the Wasatch Front, the Rockies' western edge, before expiring in his wagon in 1847 and leaving younger folk to build God's earthly outpost in what is now Salt Lake City. There is the same sense of defiance in the state's licence plate slogan, "the Greatest Snow on Earth".

Does Utah really have the greatest snow on earth? It's not hard to make the case. Six hundred miles inland from the Sierras, the climate is continental and colder when it matters.

Storm systems approaching from the west are dried out by deserts but topped up with moisture from the Great Salt Lake before colliding with the Wasatch, where the straggle of old ski hotels known collectively as Alta claims to receive 40 feet of perfect snow a year.

Dr Jim Steenburgh, a meteorologist at the University of Utah, has even written a book called *The Greatest Snow on Earth* about his local mountains and their weather, and it's not propaganda. His argument is that even though in terms of raw depth the Wasatch range is not the snowiest on Earth, the snow that falls on it is uniquely wonderful for humans at play because the storms that bring it are "right-side up". By this he means their leading edges bring heavy snow which forms a solid base for skiers, followed by pure powder to flatter and intoxicate them. Elsewhere, he says, upside-down storms are more frequent.

I asked Steenburgh in an email to put aside clan and professional loyalties and tell me where he really thought the greatest snow on earth was to be found. He obliged. "In my view," he wrote with supreme detachment, the mountains of north-west Japan provide "the greatest snow climate in the world, because the snow, at least in our current climate, is high quality, and it comes with remarkable intensity in December, January, and February during the Asian Winter Monsoon. In January, the Sukayu Onsen observing site on Northern Honshu averages

459 centimetres of snow. I suspect this is the greatest average monthly snowfall reported by any regular observing site in the world."

Sooner or later, I have to go to Japan.

CHAPTER TWO: IF POLAR BEARS COULD TALK

"The heavens are weeping for Isaiah… the universe is pulling a comforter over him so that he will never be cold again."

Peter Høeg, *Smilla's Sense of Snow*

Snow is a little like magnetic tape in a cassette – a store of information from which stories can be spun. When it falls it either melts or settles. If it melts the atmospheric data stored on each flake in the form of its shape is lost for ever. But if it settles in very cold conditions, that shape can last a surprisingly long time, and so can the stories it tells.

Freshly fallen snow is 90 per cent air. In intense cold, the branches of each flake hold their form. Since the odds against flakes falling flat on top of each other are high, they create air pockets as they settle. Over the course of perhaps two winters the weight of new snow presses down on them. Buried flakes become more granular and the snow becomes denser: the air share falls to about half.

In the next step what was once snow stabilises as firn, a snow-ice hybrid with only 20–30 per cent air. But its metamorphosis is still not complete. If the cold persists, firn solidifies as glacial ice. That process can take a century. By then the snow's usefulness as an information store is degraded but not shot. Air bubbles in this ice are what bring scientists back year after year to Antarctica. They dig it up and measure its carbon content as a way of gauging how hot or cold past climates were compared with ours.

There is a third way snow stores information. This is on its surface – in the impressions left there by wind and weather, dust, pollution, radiation – and by life. If the snow's surface stays below freezing and is not overlaid by fresh falls, this sort of information can survive for months.

In the summer of 2014 Steve Berry, a British mountaineer, was trekking in northern Bhutan at nearly 18,000 feet. At that altitude it is almost always below freezing. Berry was with a local guide on the slopes of Gangkar Punsum, the highest unclimbed mountain on the planet. They were above the snowline but below the summit pyramid, from which Berry had been turned back by a ferocious snowstorm as a much younger man, on an expedition he arranged himself in 1986. On this day nearly 30 years later the sky was clear. Around lunchtime the guide stopped suddenly and pointed to the far side of a rocky cwm. He had seen a set of footprints crossing a steep snow slope. They were neat and regular, climbing at a gentle gradient as if whatever made them was taking care not to slip. A slip would have been costly: below the snow, cliffs fell vertically to the valley. It was impossible to get close to the prints except with a long camera lens, and "absolutely impossible", Steve said, for a human to have made them. He is convinced to this day that they were made by an ape, and probably an ape that lives nowhere else on Earth.

To understand what was going on that day on Gangkar Punsum it's worth briefly going back a few hundred millennia.

The first sentient creature to see snow was probably a fish. A coelacanth, perhaps, or maybe a giant piranha or xiphactinus. In the great age of fish between the Cryogenian ice age and the dinosaurs it could have been one of thousands of fish types, most now extinct. If there were huge snowfalls as the Earth warmed up after being covered in ice, it is hard to imagine fish being completely unaware of them. Some experts believe there was once near the equator sea ice so clear that enough light could penetrate it to sustain life even when it was several metres thick. If so, something must have registered in the subconscious of the fish below as the snow fell and the light dimmed. At the very least a lungfish hauling itself onto an Aleutian mudbank sometime in the early Devonian (say, 400 million years ago) must have been startled by the feel of snow on its skin.

Time marches on. Four-legged creatures start to stalk the land. Even so, the idea of peering up at snow from underwater remains relevant to the story of snow, and its extraordinary power to shape the life of living things. From about 150,000 years ago, maybe more, the view of snow from the sea was, for ringed seals, a matter of life and death. What the seal hopes to see is a hole in the pack ice and a clear disc of blue or white above, depending on the weather. Both are welcome sights. A hole represents an opportunity to breathe. A big hole might allow some socialising. But if the disc is not clear, there may be danger. What the seal needs to look out for is a pattern of three black dots above the hole, close to the edge. For the unwary those dots – two eyes and a snout – may be the last things they see before being scooped from the water by a five-clawed paw as wide as a human's forearm is long. The claws will hurt like hell, puncturing the skin so that even an escape leaves terrible wounds. Assuming there is no escape, the seal is slapped onto the snow. The teeth behind the snout bring a merciful quick end: four banks of molars crush the seal's skull. Then, evisceration.

Such is the imperfect but deadly stealth of the polar bear. It won't be much consolation to the seal, but just as it has been tricked by the bear, the bear has been tricked by snow. There is intense debate about when polar bears split from grizzlies to form their own branch of the evolutionary tree, but there is broad agreement that between 150,000 and 100,000 years ago, grizzlies were lured into a great northward migration as the Earth warmed up. Then they were forced to learn to live with endless snow as it cooled down again.

As temperatures rose, what is now Canada discovered seasons. Greenland's ice cap shrank and coniferous forests grew up along its fjords. A whole new zone of bear country was created and colonised, not just by bears but by hares, wolves, foxes, terns and owls. They moved into what had been a vast polar desert, and would be again. They were sucked into it as Napoleon's army was sucked into Russia in the winter of 1812, but they fared better in the long run.

When the glacial see-saw changed direction and ice started to spread south again, these bears' descendants had to adapt or die. They adapted fast. By one estimate it took just 20,000 years, or 1,762 generations at 11.35 years per generation, for an enormous, complex land-based forager to morph into an even bigger beast more comfortable on ice and in the water. It became completely carnivorous, impervious to intense cold, and it turned white.

If polar bears could talk, their oral history would tell of snowstorms we can barely imagine; storms that shaped their landscape, coloured it and commanded absolute respect. Over those 20,000 years their females learned to find soft snow banks in the lee of the prevailing wind and dig deep into them. They learned to hollow out chambers big enough for themselves and at least two cubs, with upward-sloping entrance tunnels to trap the warm air they created. They learned that if digging brought them to rock or permafrost, they could, being polar bears, simply lie down in the snow and let it form a den around them. The restless drifting of dry snow would fill every crevice and make a perfect fit. Thanks to natural selection it would also be snug. The cold had selected for every bear an inner coat of soft, short fur to retain its body heat. The whiteness of the snow had selected an outer layer of hollow guard hairs up to 15 centimetres long. This was mainly for camouflage but it also served as a wraparound sleeping mat for the long polar night.

The guard hairs are not in fact white, which is appropriate because snow isn't either. Both are translucent to visible light, which they scatter in all directions, but the effect is the same: they look white, and this sets the tone of the entire Arctic colour palette.*

---

* Shine ultraviolet light at a polar bear, or look at it through a UV filter, and it tells a different story. It is not camouflaged at all. The bear looks

Polar bears and snow leopards turned white to hide from their prey. Smaller snowbound animals – and the beluga whale, which surfaces to breathe in holes in the pack ice – turned white to hide from their predators.

One snow creature didn't bother to turn white at all. It was not seen by a westerner until 1925 and even then the person who saw it, a Greek photographer on assignment in the Himalayas with the Royal Geographical Society, failed to take a picture of it. He did describe it in some detail, though: "The figure in outline was exactly like a human," Nicholas Tombazi wrote. It was "walking upright and stopping occasionally to uproot or pull at some dwarf rhododendron bushes. It showed up dark against the snow, and as far as I could make out wore no clothes." Tombazi thought he'd seen a hermit. Others decided he had seen a yeti.

Wherever the air is thin and the snow lingers, the legend of the yeti lingers too. Every few years science debunks it, and another posse of romantics heads for the mountains and becomes obsessed. These obsessions are contagious.

The closest I ever came to seeing a yeti was on a military road across the High Pamirs in 1992, when a Russian soldier pointed north to the snows of Lenin Peak, impossibly remote and pure,

---

dark against a snowy background. For at least two decades several groups of scientists propagated the myth that this was because a polar bear's guard hairs acted as fibre-optic tubes, channelling UV light to its dark skin to warm it. Then Professor Daniel Koon of St Lawrence University, New York, took some measurements and found that only about 1 per cent of the available UV light was transmitted any distance by the hairs. Most of the rest was absorbed by them because, like horse hair, they are made of keratin, and keratin absorbs UV.

and told me as if talking about a type of rabbit that the *snezhnyi chelovek* lived there. There was no doubt in his mind, and therefore no need to go and check.

The *snezhnyi chelovek*. The snowy fellow. The abominable snowman. Back then, anything seemed possible. A few days later I must have managed to convey some of the soldier's conviction in a phone call to a now defunct newspaper in London. It probably helped that Lenin Peak had been sealed off from the outside world for 70 years but in any case, for a few pounds, I topped up the story of the yeti for a readership thirsting for news of curiosities preserved by the Soviet Union.

The status of the yeti has since taken something of a beating. It's now on a par with that of UFOs and strictly for the tabloids, but at the time people still welcomed any excuse to suspend disbelief. Part of the reason may have been the miraculous collapse of communism, but another part was snow – a square foot of it that 40 years earlier had received an imprint that went around the world; an imprint that launched a thousand newspaper articles even more excited than mine.

This patch of snow was on level ice towards the foot of a long glacier north west of Mount Everest. It was between 15,000 and 16,000 feet up, two to four inches deep on a base of firn, and something had stepped on it. Whatever it was had left a footprint 13 inches long and nearly as wide. It was bulbous and cartoonish, with an imprint the diameter of a tennis ball where its big toe would have been. And it was one of a set. Its creator had been minding its own business, walking down the glacier.

No human would have seen it if Eric Shipton hadn't walked

that way a short while later. Shipton was a climber on a British Everest reconnaissance expedition, with time on his hands before returning to Kathmandu and London at the end of the season. He laid the head of his ice axe down next to the print to show its scale, and took a picture that was first published on the front page of the *Daily Mail*. It was immediately picked up by hundreds of other newspapers that would have loved the scoop, and with good reason.

The Shipton footprint has the power to transport. Even now it takes you in the blink of an eye to a place of mystery and monsters and adventure. The mystery is the print. The adventure is in the axe. The monster is somewhere nearby, possibly behind you, real enough to set the mind racing. All by itself that picture made the yeti a suitable subject for polite, if breathless, conversation for the next two generations.

There are many theories about what made the footprint. According to one it was a hoax, perpetrated by Shipton in frustration at having been passed over for leadership of the 1953 British Everest expedition. But it was not a hoax. The footprint was photographed more than a year before John Hunt was named the leader. Other theories were later offered by Michael Ward, a young doctor who was on the glacier with Shipton that day and also saw the prints. He said they could have been made by a bear, a langur monkey or a human with deformed feet. In an article for the *Alpine Journal* he helpfully included a picture of a Nepalese highlander's feet with big toes pointing towards each other at right angles to his other toes; and the story of Man Bahadur, a Nepalese pilgrim who spent two weeks above 15,000 feet in the Everest region in the winter of 1960–61, walking shoeless over snow and rock at minus 15°C with no frostbite or other ill effects.

The point of this story was to remind western readers in their slippers that people can walk barefoot in the snow without complaining. Add to this the fact that footprints in snow get bigger by melting outwards, and it became easier to account for what Shipton had seen without reaching for the yeti. But where was the fun in that?

There was another explanation: a snow leopard had made the tracks by planting its hind paws neatly in the impressions left by its front ones. This was exotic but fanciful. Neither Shipton nor Ward had even considered it. The yeti explanation has dominated over the years because it's hard to prove a negative, and because people like it. In the late 1950s they rushed to read Hergé's immortal *Tintin in Tibet*, the story of a boy lost in the Himalayas after a plane crash and rescued by a lonely yeti thrilled to have company. And in the late 1980s two of the world's best-known mountaineers went to Tibet to look for him.

For Chris Bonington and Reinhold Messner, nothing else would have warranted such a long and expensive distraction from the business of reaching summits, but the yeti was worth it. Bonington was the high priest of British Himalayan siege climbing; a leader, hard as nails but smooth on the outside. Messner was the lone wolf from the Dolomites who climbed 8,000 metre mountains without support or oxygen and made it look easy. Both were captivated by the yeti and madly competitive about him.

Both undertook expeditions in 1988. They crossed paths in

June that year in the Lhasa Hotel in the Tibetan capital, where a third climber, the American Ed Webster, was resting after an arduous attempt on Everest. Webster wrote:

> When we met him, Messner said mysteriously: "You should not even ask me why I am here." So we didn't. Yet soon it became evident that, coincidentally or not, Messner and Bonington were each seeking something considerably more elusive than the summit of an untrod Himalayan peak [...]
>
> "Where have you been climbing," Reinhold asked Chris, in an understated yet measured tone of voice, his piercing blue eyes focused and intent.
>
> "Um, on Menlungtse," Chris admitted.
>
> Messner examined every nuance of Bonington's answer. "Ah, yes," Reinhold said. "That is a very good place to see the yeti." Bonington smiled.
>
> "What do you think the yeti looks like?" Chris wanted to know.
>
> Reinhold puzzled over this for a minute. "I have a very good idea of how the yeti appears. He has light grey or brownish fur. He can walk on all fours or upright. He eats fruit and vegetation, but also meat on occasion... [and] sometimes he goes walking on glaciers."
>
> "Then you're convinced that they're real?" Chris persisted.
>
> "Yes, of course they are real," Reinhold answered, then paused slightly before exclaiming: *"I have seen one!"*

The great thing about being Reinhold Messner is that no one can tell you you are wrong. He has been to so many high places as a solo climber that he has to be taken at his word or not taken seriously at all. Either way, neither he nor Bonington found any proven trace of the yeti that year. In 2014 the Oxford geneticist Bryan Sykes produced a book based on tests of dozens of unproven traces consisting of skin and bone and hair. None came from an ape or a biped of any kind. One, tantalisingly, seemed to share DNA with the world's oldest known polar bear remains, a 130,000-year-old jawbone found in Svalbard. This raised the possibility of a trans-continental polar bear migration back from the Arctic towards the lower latitudes, but no one else has been able to replicate Sykes's results.

Steve Berry is not particularly bothered. He knows what he saw high on Gangkar Punsum that same year; the gently rising footprints on the snow no human could have reached. Four days later he descended to the valley and was told matter-of-factly he had seen the migoi – Bhutan's yeti. He went back in 2016 and found another set of prints, these ones zigzagging upward over a mixed slope of snow and rock. Whatever made them, he said over the phone while I looked at the pictures on a computer screen, "you can see that it's creeping up on something".

"It's going really quite slowly. It gets to this point where it turns on one foot, then strides on much faster, like it's stalking something. I can't imagine a four-legged animal turning like that, so I'm sure it's an upright creature and I'm sure it wasn't a bear."

Bears walk upright, he admitted, "but they don't put one foot exactly in front of the other". In the pictures, one foot exactly in front of the other is how they were, and Steve was confident that the valleys east of the mountain were big and undisturbed enough to hide a migoi. "I think it was some kind of ape," he said. After talking to him I emailed the world's leading expert on the ancient Svalbard polar bear jawbone, just to be sure there was no chance he'd seen the footprints of a hitherto undiscovered polar-Himalayan hybrid.

"I think it is theoretically impossible that polar bears *as we know them today* have interbred with Central Asian bears," Dr Charlotte Lindqvist of the University of Buffalo replied (her italics). "But it is not unlikely that their ancestors could have interbred and that some genetic material could have been carried over to today's populations and left traces of past interbreeding."

She was much more open to a Sykesian hybrid scenario than I had expected. If polar bears could talk they might be able to solve the mystery of the yeti once and for all, but they can't, which leaves only the snow, and the snow tells whatever story its beholder wants to hear.

CHAPTER THREE: SNOMO SAPIENS

"As though there were not enough to worry us already. This snow—"
"Snow?"
"But yes, Monsieur. Monsieur has not noticed? The train has stopped. We have run into a snowdrift. Heaven knows how long we shall be here."

Agatha Christie, *Murder on the Orient Express*

On February 10th, 2018, Simen Krueger woke up nervous. He was young and fit but a long way from home. More to the point, he faced a test of endurance so extreme that the mere thought of it was enough to make some participants throw up.

The 30-kilometre-cross-country skiathlon is a brutal way to get your exercise. It is even more painful as a route to an Olympic medal. The first 15 kilometres have to be skied in the classic style, skis parallel, arms and legs pumping in an elongated version of running. The second 15 are freestyle, which in practice means a skating motion but with two-metre skis instead of skates. As anyone who has tried this knows, the skis dangle from the toes of lightweight boots, inviting you to trip at every step.

Krueger was 24 at the time; one of the younger members of

a Norwegian winter Olympic team that had arrived in Pyeongchang in South Korea with the benefit of year-round Arctic training, but also with the expectations of a snow-mad nation weighing on their shoulders.

Cross-country skiing was invented in Norway. It was honed in Norway as an essential method of transport and survival and used in Norway by resistance fighters against Swedes in the tenth century and Nazis in the twentieth. It is Norwegian nationhood in motion. The silhouettes it makes of lithe human forms against the snowscapes of the north are Norway's pride and joy, and while most Norwegians recoil from jingoism they do expect great things from their Olympians. In the skiathlon they wanted gold, silver and bronze.

For Krueger, disaster struck on lap one. A hundred metres into the race he slipped and fell, tripping up two Russians who fell on top of him. The tangle of six skis, six poles and 12 limbs took 40 seconds to sort out, leaving all three far behind the back of the pack. Krueger realised one of his poles had snapped into the bargain. He was a four-cylinder engine firing on three cylinders. He set off anyway to start making up the lost ground.

As he said afterwards: "I was completely last."

Poles break surprisingly often in big cross-country races and competitors are allowed to accept replacements, which Krueger did. Even so, he was still 37.8 seconds off the pace at the six kilometre mark. Ordinarily he would have needed to be close to the leaders at that point to have a realistic chance of a medal because when a frontrunner makes his move a challenger needs to be able to go with him. Instead, Krueger was just beginning to work his way back into contention. To get to the front he would have to overtake 63 other competitors. They were not taking it easy – they were the best in the world and this was the Olympics – but after a few more laps he found himself enjoying the unfamiliar challenge of coming from behind.

"OK," he'd said to himself, "take one lap, two laps, three laps, and just get into it again."

One after the other, Krueger reeled in the field. After 24 kilometres he was where he needed to be, just behind the leaders, with the whole of Norway cheering him on. He made his move with five kilometres to go, opening up an eight-second gap which two fellow Norwegians protected for him, skating in echelon so that anyone with the strength to take him on would have to get round them first. No one could, and Norway won the sweep.

The race recalled another that is etched in Norway's consciousness. It, too, involved a comeback and a broken pole. In February 1982 a popular skier who had never quite realised the potential his admirers saw in him was given the anchor leg in the men's 4x10 kilometre relay at the Nordic Ski World Championships. His name was Oddvar Brå. The race was staged before a huge home crowd in Oslo and Brå started his leg 15 seconds behind the Russian team. The gap had opened up because the skier before him had had a freak fall on a downhill section for which he later received with death threats.

Brå started well, closing the gap. On the final lap he was poised to overtake the leading Russian on a steep uphill section. At 1.54pm on February 25th, he did. The time is remembered every year when the footage is replayed on national television. The reason: it was not a clean overtake. The two men touched shoulders. Whose fault it was has never been agreed but the Russian, Alexander Zavyalov, fell. And Brå broke his pole. The commentator screamed for someone to give him another. Zavyalov picked himself up and mounted a mini-comeback of his own. They crossed the line neck and neck and after an hour of deliberation the judges gave them joint first place.

"Where were you when the pole broke?" has been a Norwegian conversation starter ever since, and 26 years after the event the *New York Times* retold Brå's story under a photograph printed across six columns of him skiing in the classic style, aged 66, through a forest daubed in heavy snow. The picture makes a powerful impression, especially when you consider the legend behind it and Norway's utter dominance in its national sport: 111 Olympic medals over the years, including 15 gold-silver-bronze sweeps. You gaze at Brå, deep in that forest, and wonder if you might actually be looking at a new sub-species; if Norwegians are evolutionarily adapted for snow, like polar bears.

The idea is not completely mad – most geneticists concede our species is still evolving – but that doesn't make it true. And if it *were* true it would still not account for Norwegians' success in Nordic skiing, because the same adaptation would have to be found among Russians, Swedes, Canadians and Finns.

So why don't these other Arctic nations win as much? The answer seems to involve Norway's history and culture rather than genetics. For no other country has embraced with quite so much energy and interest the slipperiness of snow.

Why is snow slippery? This is two questions in one. The first is philosophical – to what end? – and the answer is: to none. Unless you see an unmoved mover spinning the planets and firing up the stars it's hard to see a purpose in the slipperiness of snow. It just is slippery, thank heaven. It's pure good fortune that a ski base or toboggan runner will slide over snow with such blissful ease when pressed down and helped along by gravity or muscles.

The second question is "how come…?" and herein lies a tale of molecular-scale science almost as miraculous as the one that describes the formation of snowflakes in the first place.

Not by chance, some of the most meticulous work on the "how" of snow's slipperiness has been done in Norway by a ski wax firm. Those Norwegians know they cannot trust genetics. To win, they have to get their wax right. In a 2005 paper titled "Why is Ice and Snow Slippery? The Tribo-physics of Skiing", three engineers hired by the Swix wax company of Lillehammer pulled together more than 60 years of research to explain what happens when something slides over snow, and to answer a follow-up question that had defeated many of their forebears.

Tribology is the study of friction, and sliding on snow happens when friction is overcome. The basic explanation is simple: a combination of pressure, friction and freely available water molecules creates a thin fluid film between an object (a ski, say) and snow. This film means much less friction than two dry surfaces rubbing against each other, and much less drag than with a thicker fluid layer. The high friction between dry surfaces is easy to intuit. The high drag from too much water is less easy to picture. Water-skiers seem to overcome it, after all. But they do so only with the help of powerful outboard motors. Water does drag, and anyone who has been skiing in rapidly warming conditions is likely to have experienced the strangeness of snow that seems to be grabbing at their skis. Hydrodynamicists call this capillary drag, and when you expect to feel the wind in your hair it isn't nice.

In 1902 a German mechanical engineer, Richard Stribeck, put each of these three degrees of lubrication – none, a little and too much – on the x axis of a graph, from dry to wet, and plotted net friction for each of them on the y axis. The result was what became known as the Stribeck curve, dipping to a sweet spot of minimum friction created by a Goldilocks amount of mixed lubrication.

In snow this lubrication consists of water, and this is the reason for the follow-up question addressed by Team Swix: in very cold conditions, why doesn't this water freeze solid and bring skiers to a halt? In the late 1930s a Tasmanian-born scientist at Cambridge, Frank Browden, tried to answer this by hauling a specially built simulator 3,300 metres up to the Jungfraujoch research station in Switzerland to study snow surface melting as a result of pressure and friction down to minus 20°C. He decided that the lower the temperature, the greater the role played by "frictional heating" in keeping things sliding. There was a paradox, in other words – friction fighting friction – but he failed to explain it satisfactorily.

In 2005 the Swix team explained it better. By then the science of super-cooled liquids was well understood. In the absence of dust, water can stay liquid to minus 40°C. Water molecules need that dust to "nucleate" around and fall as snow, but having done so they don't need anything to reverse the process. They don't need any equivalent encouragement to detach themselves from the orderly lattice of an ice or snow crystal or at least form an invisible boundary layer of looser molecules with all the slippery properties of water, but none of the capillary drag. As Team Swix wrote: "The reason why ice and snow is slippery over a wide temperature range lies in the fact that there is no nucleation barrier for melting… For skiers and others who wish to explore the fun and beauty of snow activity, this temperature range fortunately coincides with our normal winter temperatures."

This happy coincidence of sliding temperatures and winter temperatures has held for thousands of years. Norway's museum of cultural history in Oslo has wooden skis carbon-dated to 5000 BC. If that seems a long time ago for an activity normally associated with late capitalist hedonism and clanking electric lifts, consider how long *Homo sapiens* must have been aware of snow.

The words "out of Africa" sound like hard work. They evoke a long, hot walk from the Rift Valley to Sinai and Eurasia; a baptism of endurance for early humans. But there's another way of looking at it: Arcadia to Arcadia. It started within sight of snow and ended in the midst of it. We have to discipline ourselves to remember that the great migration out of the cradle of mankind wasn't a journey. It took hundreds of generations, each one moving on only when a cost benefit analysis of the visible savannah versus whatever lay over the horizon persuaded the alphas of the group that it was worth it. Even so, I like the idea of a small group of Neanderthals or early *Homo sapiens* striding out, sandwiches in knotted handkerchiefs, looking fondly at the snows of Kilimanjaro to the east, waving farewell and vowing that their children's children would one day return. Perhaps they had started even further south, in the foothills of the Mountains of the Moon, intending to follow the flowing meltwater of the glaciers on what is now the Ugandan–Congolese border wherever they led. Either way their descendants ended up 20,000 kilometres away in a different set of foothills, in the lee of the mighty Chugach, where snow and ice reach down from the heavens to a land of deep water, dark forests and natural abundance.

These were the Amerindians who crossed the Bering Land Bridge around 14,000 years ago and turned right on reaching North America. They followed the coast south-east to what is now Alaska's inside passage. Ultimately their descendants continued to Patagonia. Others stayed in the far north. For both groups snow was part of life and had been for millennia.

How many millennia? One answer offered recently is 45.

The first human farewell to Kilimanjaro occurred anything from 185,000 to 60,000 years ago, depending on your definition of human. As those early hominids drifted north up the Nile they would have been starved of the sight of snow for several thousand years. They would have been reacquainted with it in what is now Lebanon, but only as something that came and went on high ground, left there by clouds and melted by the sun. They would have seen it in the Bekaa Valley and again on Mount Ararat. They would have coped with it all over Europe, but only as a seasonal variation.

The most fearless migrants among them eventually made it

to the Arctic. This was humanity's first exposure to snow as a defining feature of the landscape and it started at least 45,000 years ago. The evidence takes the form of hunting marks on mammoth bones in northern Siberia. The oldest so far found were dug from a bluff overlooking the mouth of the Yenisey River about 1,000 miles north of the Arctic Circle, in 2012. Their great age pushed back the date of the first human presence on the permafrost by 10,000 years. If the oldest known skis are from 5000 BC that makes them 7,000 years old. Subtract 7,000 from 45,000 and you are left with 38,000 years for humans to discover the slipperiness of snow. Viewed like that, those skis seem positively modern, even though they are twice as old as the pyramids. Nor does it seem so odd that there was a Norse god of skiing before there was a Greek god of war.

In practice, of course, our species must have started sliding on snow long before it first strapped planks to its feet. To experience snow as a big white banana skin all you have to do is fall over on a slope. From that it was surely a simple step to take to control the slide instead of being ambushed by it. And so it proved.

For the opening scene of *Dr Strangelove*, Stanley Kubrick's nuclear meltdown masterpiece, he somehow obtained footage of the Soviet Zhokov Islands, shrouded in low cloud, dark mountains poking through, shot as if from the cockpit of an approaching B-52. In the film the islands are the deserted Arctic frontier of one militarised madhouse being attacked by another. In reality they were already inhabited 9,000 years ago by hunter-gatherers with dogs, and harnesses, and sleds. Remnants of all three have been analysed by Vladimir Pitulko, a palaeo-anthropologist who has probably spent more time on the Zhokov Islands than anyone alive. In 2017 he enjoyed a modest breakthrough with a widely cited paper on the dimensions of 11 fossilised dogs' skulls. They were smaller than wolves', he said, suggesting they were bred for pulling. As to what they might have pulled, by then Pitulko had been finding pieces of sleds in Siberia for nearly 30 years, and not just any sleds. He says they were two metres long with wooden uprights 30 to 40 centimetres high. They were built for loads of up to 150 kilograms excluding a driver, who might be half-on or half-off, depending on the slope.

This was so long ago that the last glacial maximum* was still receding and the Zhokov Islands were still connected to the Siberian mainland. The image of dogsledders gliding over snow between them had me wondering. Could they have crossed into America the same way? Instead of trudging to the New World, is it possible that early man slid there? I sent an email to Pitulko late one night to put the question. He was in St Petersburg, three hours ahead of GMT. He replied at length at two in the morning his time.

"The land bridge was crossable either by walking or by using sleds if they had them about 14,000-12,000 years ago," he wrote. "Although there is no direct evidence of sleds at that time, technically it was possible. Why not?"

No *direct* evidence. So what about indirect? Pitulko had

---

* Peaking around 30,000 years ago, the last glacial maximum covered much of what is now Canada, Siberia and Tibet in huge ice sheets. These stored so much water that according to the US Geological Survey global sea levels fell by 125 metres, enough to expose huge areas of former sea bed including the Bering Land Bridge.

plenty. First he seized on the idea of the human discovery of the slipperiness of snow, which was gratifying since as far as I knew no one else had. Then he plunged back 27,000 years to another stash of mammoth bones. This time they were on the Yana River at 71 degrees north, two-thirds of the way from Moscow to Alaska. The site was not properly excavated until he and others got to it in 2001. They quickly realised this was not a place where mammoths came to die. Nor was it the site of a massacre.

"It was an accumulation that took at least five to six thousand years," Pitulko said. "It was very slow – perhaps one or two animals a year, and the bones were sorted. Only humans could have done that sorting, so the bones had been moved to that place by humans, and this would have been a heavy load to move across the tundra." Pitulko was posing the Stonehenge question – how did they do it? His answer: "I think it should be easier in winter by using skins to drag the stuff. I think that is how it was."

Ice road truckers, eat your heart out. Snow-aided winter haulage started 27,000 years ago, when ice sheets still covered much of the northern hemisphere. Dragging mammoth tusks on skins admittedly sounds primitive, but on the Siberian mainland those pioneers may have graduated to sleds even earlier than on the Zhokov Islands. As ever, the case rests on the never-ending search for food, in this case reindeer meat.

"As it is a migrating animal you need to move often and fast just to keep up," Pitulko explained. By around 15,000 years ago wild reindeer were Eurasian humans' most hunted animal, with a migration route 500-600 kilometres long each way. "And that made the hunters think of dogs and sleds. The high level of development of Zhokov sled technology suggests some time had already been spent on experiments and learning, so I think the process would have started about 15,000 to 13,000 years ago and then spread across Siberia."

That was good enough for me. What applied to Siberia surely applied equally to Beringia, the now-submerged lowland between east Asia and Alaska. I went to sleep that night with visions of dog sleds sweeping across the international dateline bearing the true discoverers of America, 16,000 years before the Vikings and 16 and a half before Columbus. Truly, snow shaped history.

The descendants of the first humans to cross the Bering Land Bridge settled two whole continents. All were formed by their surroundings, but none so dramatically as the Inuits. They adapted their hunting techniques to target seals and whales in blowholes in pack ice. They adapted their metabolism to run hot in extreme cold on a diet of meat, fat, calories and almost no fibre. And they adapted their language, as the *New York Times* stated in an editorial in 1984, to distinguish between "one hundred types of snow".

The mention of words for snow will make some readers' hearts sink. The idea that Inuits or Eskimos (or both) have dozens or hundreds of different words for snow (or no more than anyone else) has for more than 30 years been a bone of bilious contention between two groups of tetchy academics – those who have first-hand knowledge of Inuit/Eskimo linguistics, and those who pretend to.

The story is worth recounting briefly, not for what it says about Eskimos but for what it says about the rest of us.

It starts in 1986, with publication in the journal *American Anthropology* of a paper by Dr Laura Martin of Cleveland State University titled "Eskimo Words for Snow: A case study in the genesis and decay of an anthropological example". Not very gently, Martin chides a parade of her peers for peddling the idea of multiple Eskimo words for snow on little evidence. What evidence there is comes from Franz Boas, a German-born anthropologist who travelled to Baffin Island in the 1880s. He delighted in the company and culture of the Inuits. He wrote to his fiancée that he was living among them "entirely on seal meat", and he came back with four Inuit words for snow: *aput*, *qana*, *piqsirpoq* and *qimuqsuq*. These meant "snow on the ground", "falling snow", "drifting snow" and "snow drift", respectively.

Boas mentions these in the introduction to his *Handbook of American Indian Languages* in 1911. Three decades later a self-taught linguistics enthusiast from Connecticut, Benjamin Whorf, has an essay published in the Massachusetts Institute of Technology's *Technology Review* (not because it's an appropriate journal, which it isn't, but because he has an in, being a graduate of MIT). In the essay Whorf raises the number of Eskimo words for snow to at least seven, without listing any sources. Boas' point was to draw attention to the similarities between Eskimo and English. Whorf's point is to draw attention to the differences, and this starts something. Others see his seven and raise it to 50, 100, 200 and beyond. No one is counting or caring any more until Martin calls time on an "object lesson on the hazards of superficial scholarship".

For a couple of years few people notice Martin's brave challenge to received wisdom. Then Geoffrey Pullum, a British linguist and former rock-and-roll pianist at the University of California in Santa Cruz, decides to play the Eskimo snow words debate for laughs. In 1989 he produces a paper called "The Great Eskimo Vocabulary Hoax" that likens the "self-generating myth of Eskimo snow terminology" to the monster in *Aliens*. He ventures that most university linguistics teachers will at some time or other have told students a story about "these lexically profligate hyperborean nomads". Then he says: "What a pity the story is unredeemed piffle."

Pullum's paper goes the 1989 equivalent of viral and suddenly the orthodoxy that Eskimos have a lot of words for snow is replaced with a new one: they don't. They have a normal number that can be made to look large because of the way their dialects are structured, with multiple suffixes added to a few root words to make an almost infinite number of long and complicated compound words. Pullum says the public's willingness to believe anything about Eskimos is buried racism. He is especially brutal in his treatment of Whorf, the "Connecticut fire prevention inspector [which he was] and weekend language-fancier", for being variously glib, wrong and, worst of all, "quoted and reprinted in more subsequent books than you could shake a flamethrower at".

This all feels a little unfair, partly because Whorf died of cancer in 1941 and was never able to defend himself, and partly because Pullum, in the end, is the one who gets it wrong.

In his initial "hoax" paper, he commits the sin of which he and Martin accuse others. He consults no Eskimos. He is honest enough to see the problem with this approach, and by the time the paper is reprinted in a book two years later an appendix is added by an specialist on Central Alaskan Yupik, one of five main Eskimo dialects. The expert, Professor Anthony Woodbury, lists 15 root words for snow from an authoritative Yupik dictionary. This is not hundreds, but it's many more than either Boas or Whorf claimed. They are all distinct. They don't incorporate a general word for snow as English often does (for example in snowflake, snowdrift or snowstorm) and they don't have English equivalents. In addition to Yupik versions of the four snow words listed by Boas, Woodbury offers *kanevvluk* (fine snow), *muruaneq* (soft, deep snow on the ground), *navcaq* (snow cornice or formation about to collapse) and *qanisqineq* (snow floating on water). Each, says Woodbury, may have about 280 distinct inflected forms.

Whatever Pullum intends this to mean it doesn't seem to support his view that the idea of many Eskimo words for snow is "unredeemed piffle". For what it's worth, a team from the Smithsonian Arctic Studies Center piled into the debate more recently. They counted 53 distinct words for snow used by Inuits in Canada's Nunavik region, and 40 in Central Siberian Yupik. There are still plenty of academics and fashionable journalists on the other side of the argument, but they tend to have in common a lack of first-hand experience or specialist knowledge of the Arctic. Some go there, of course, but I have a theory that they may undercount the number of local words for snow when they do. This is because they go mainly in summer, when there isn't any snow to talk about. This theory is completely unscientific and I am the only source for it, but I've been to the Arctic four times and never seen or talked about a single snowflake there.

Nearly three decades after writing his original essay, Pullum, now a professor at the University of Edinburgh, is still keen to clarify its purpose. "It's not Eskimology," he explains in an email. "The tribes I was focusing on were not the Yup'ik of west Alaska or the Inuit of Kalaallit Nunaat, but the anthropological linguists of the USA and the teachers of Linguistics 101." Which is fair enough, but it doesn't mean the argument over the number of Eskimo words for snow has been settled in his favour.

Followers of the Whorfian view of Eskimo snow meanwhile stumble on two points. They fail to recognise that just like the Eskimo dialects, English has plenty of words for snow, both falling (flakes, crystals, graupel and diamond dust, not to mention the dozen or so more technical flake terms on the Nakaya Diagram); and fallen (powder, slab, windpack, slush, sastrugi). And they pick a race rather than a specialism such as, say, car mechanics to make the uncontroversial point that people learn to describe in greatest detail what matters most to them.* That's why Pullum can accuse them of harbouring buried racist tendencies.

* Granted, it's more complicated than that. Whorf suggested that Eskimos' big snow vocabulary enables them to think about snow in more sophisticated ways than the rest of us. This view of language has been largely debunked by those who believe words follow thoughts rather than vice versa, but this isn't a book about linguistics.

But surely there's another way of looking at this. Whorfers aren't racist. They're just fascinated by snow. They like the idea of needing 200 words for it even if not all those words exist. There is an echo of the yeti here; an expectation that in the realm of snow things will be different and mysterious. But there is a difference, too: unlike the yeti some Eskimo words for snow do probably exist, and one of my favourite is *utvak*.

Utvak. Two uncompromising syllables preceded by a firm glottal stop. If it sounds like a building material, that's because it is. Woodbury gives its meaning as "snow carved in a block". The technical term for it in English would be *sintered* snow, but there's a better way of saying what it really is. It's igloo snow.

Dry powder is useless for snowballs, never mind igloos. Snow that has been stomped on even once is different. The stomp breaks the branches of dendritic snowflakes and bonds together the simpler, stronger shapes that are left. Snowmobiles and piste grooming machines have the same effect. So do gravity and wind, over a longer period of time. They all create a new material with low tensile strength but high compressive strength; a material that can be sawn into regular blocks, squeaking like polystyrene, and lifted from the snowpack like lightweight concrete. You can sit on a block and not collapse it. You should be able to stand on top of a well-built igloo. Having built it, you can use more sintered snow to furnish it with beds, benches and coffee tables.

Igloos are heavily romanticised but they aren't comfortable.

What they are is an elegant fusion of physics and environmental living. Being roughly hemispherical, they maximise space for a given amount of snow. They have no corners and no wasted space. They can withstand any blizzard since the effect of drifting snow is to make their walls thicker and their shape more aerodynamic. Even without the benefit of storms they get ever stronger as long as the weather stays cold, as the blocks used to build them fuse into a single unit. This snow still contains air bubbles, making it a natural insulator. When it's minus 30°C outside it can be zero or warmer inside with a little help from the low flame of a kudlik of seal blubber. When people are inside, breathing and otherwise creating warmth, a thin film of water forms on the inside of the igloo. When they're out hunting it freezes into an airtight skin.

The biggest igloo ever was built by a team from Volvo on a ledge beneath the Matterhorn. That took three weeks and 18 people and lasted two months. It was 12 metres wide and nearly as tall, but the beauty of traditional igloos is that they can be built in 40 minutes. The accommodation will initially be cramped, but the base can be dug out to create a snow Tardis far more spacious than it looks from outside.

The classic igloo is an inward-leaning spiral of snow blocks. The blocks should be rectangular from the outside but wedge-shaped in cross-section to create the lean, and they should get smaller as the spiral rises. Inuits used to cut them out of the windpack with sharpened caribou bone or narwhal horn. In the 20th century machete-like snow knives made of steel took over, and now saws are made for the purpose.

Expert builders can cut all the blocks needed for an igloo

from a single neat trench. Often this becomes the basement; a cold sump with tiered sleeping ledges above and an entrance tunnel at ground level with a right angle to keep out the wind. The most important skill for weatherproof construction is the ability to shape each new block to ensure a snug fit with those below it and the last one in the spiral. This is done with a snow knife or saw, and it means the igloo is built anti-clockwise if the builder is right-handed; clockwise if left.

Watching the process is mesmerising. You can do it over and over again courtesy of the National Film Board of Canada's *How to Build an Igloo*, a 1949 masterclass in igloo-building and condescension:

> Chupak and Agiutak have left their families in the igloos of the winter post to trade. They look forward to a mug-up of tea and pilot biscuits in the trader's house. The two Eskimos admire the wooden buildings of the white man, but for their own dwelling they will build an igloo.

Since then most Eskimos have acquired wooden buildings of their own. Snowmobiles have ended the need for multi-day hunting trips and therefore for igloos as overnight shelters. But there are other uses for sintered snow; for *utvak.* Shaped by machines, it makes the half-pipes and the giant ski jumps that entertain nations with attention spans too short for the Nordic biathlon. Smoothed by the wind, it makes runways at the South Pole that can take the weight of a fully laden C-5 Galaxy transporter. Released by an avalanche, it can encase an unwary skier as if in quick-setting concrete, and suffocate him in two minutes flat.

CHAPTER FOUR: WHAT BRUEGEL SAW

"It was evening all afternoon. It was snowing..."
Wallace Stevens, "Thirteen Ways of Looking at a Blackbird"

Scanning history for great snow events, it's easy to assume the visual record is going to be the most compelling. Then you remember that even if it is, it won't be reliable. Photography is a recent invention. Painting takes a long time and there was never more than a tiny chance that an artist would be ready with easel and brushes to record a moment as an iPhone can. This is especially worth bearing in mind when bringing a detective's eye to the paintings of Joseph Mallord William Turner.

Turner was the Shakespeare of British painting, and a fibber. Late in life he cultivated a story that he had arranged to be strapped to the mast of a steamer for four hours at night to witness a snowstorm at sea. The story is told in the long title of a painting he produced in 1842: *Snow Storm – Steam-Boat off a Harbour's Mouth making Signals in Shallow Water, and going by the Lead.* The inscription added: *The Author was in this Storm on the Night the Ariel left Harwich.* Except that he almost certainly wasn't, because there was no *Ariel* off Harwich on the night in question and there is no record of him even being in the neighbourhood. Not that posterity has complained. The painting, a wild vortex of snow conveniently illuminated by the steamer's lights, was admitted swiftly into the pantheon of the greats and now hangs in the Tate Gallery. That was the second deceptively real snow painting by Turner. The first appeared nearly 30 years earlier.

In the winter of 1802 Turner was in Switzerland. He was 26 going on 27, already busy reinventing painting. "My job is to paint what I see, not what I know," he said, and he sketched furiously at every opportunity. So it would be reasonable to think that one of the most famous paintings inspired by this trip – *The Fall of an Avalanche in the Grisons* – was based on something he saw. In fact there is no evidence he saw an avalanche at all. He didn't produce the painting until eight years later and was probably inspired to by hearing about a disaster that occured in 1808.

The 1808 avalanche was a monster. It followed three days of heavy snow over most of eastern Switzerland between December 11th and 13th. A cluster of avalanches struck on the nights of the 12th and 13th and one of the most powerful thundered into Selva in the Grisons, mid way between Zurich and Locarno. It destroyed 81 barns and nine houses. It killed 24 people and 355 cows, and it made the local papers, some of whose readers would have known Turner from his travels. This seems to be how he learned of it. With memories of the mountains and descriptions of the avalanche in his head, he went to work.

The villain of the painting is a huge boulder about to crush a hut in the floor of a steep-sided valley. The boulder has tumbled down from on high, along with snow that fills much of the background. The snow is rendered in blocks, heavy-looking and hard-edged; if the boulder hadn't got the hut the snow would have. So Turner has read the reports and understood how the process of avalanching transforms snow into something solid and immovable, but he wasn't there. Nor was he there to witness Hannibal crossing the Alps in 228 BC, but that didn't stop him painting Hannibal's soldiers into yet another snowstorm of his imagination.

What is remarkable about all these paintings is that Turner even tries to depict specific snow events in which snow is an active force of nature. Most painters don't bother. Western artists in particular have generally been poor chroniclers of snow, and even when intrigued by it they have treated it mainly as a special effect. Some have been moved by it. Few have been excited. I defy anyone at Christie's or Sotheby's to point to a child throwing a snowball in a notable painting from the age of the grand masters. They were too serious for anything like that.

Some of this may come as a surprise to fans of Bruegel. Pieter Bruegel the Elder (1525–69), a grand master among grand masters, has a room to himself the size of a basketball arena in the Kunsthistorisches Museum in Vienna. It is a monumental display in every sense. The paintings are enormous and dark, in heavy frames, lowering like thunderclouds. Despite this, their creator managed to be remembered for, among other things, introducing snow to art. In reality, of course, others got there first. One of the earliest major European artworks to feature snow in a starring role is the February page of an illuminated

Gothic calendar produced for a French prince in 1416. The snow here makes a decorous coverlet. It turns surfaces white in a scene of otherwise unremarkable medieval contentment. Doors are left wide open. A piper and a lady in matching blue garments are going out of their way to suggest the weather is not troubling them. Later in the 15th century a young Albrecht Dürer took a much more daring approach to snow in a watercolour dashed off as he crossed the Italian Alps on his way home to Nuremberg. He painted late-spring snow as rolling eiderdowns that smooth the mountains' contours. This may have been the first and last attempt for hundreds of years to depict a proper dump, but the picture hangs largely unnoticed in the Ashmolean Museum in Oxford. Rightly or wrongly Bruegel is the first big name to bask in snow's reflected glamour.

His snow settled early in my mind because it was depicted in my grandmother's dining-room placemats. Every mat featured snow: snow on the roof of a 16th-century Dutch inn, snow on the curved tops of water wagons, snow beside frozen ponds, snow on the ground outside a crowded meeting house. For years I assumed Bruegel painted almost nothing but snow. Then I went to Vienna and discovered that he painted it a grand total of three times and never with the slightest interest in depicting real events. Every scene on the place mats was a detail from one work – *The Census at Bethlehem* – that plainly is not set in Bethlehem. Not one of the scenes on the mats was from the only Bruegel painting with "snow" in the title; the one that put snow on the list of subjects suitable for respectable artists to paint. This was *Hunters in the Snow*, the most famous portrait in existence of Europe in the Little Ice Age. Bruegel produced it in 1565 and it is reproduced each year by the truckload as a Christmas card.

Three men and a pack of dogs enter stage left with their meagre haul of a single fox. Nearby a roaring fire is being tended outside a pub whose sign hangs askew. Down a short slope in the middle distance a whole village seems to have come out to play on a pair of frozen ponds. Beyond, wintry lowlands stretch to the horizon while in the top right-hand corner jagged mountains rise steeply and improbably into a slate-grey sky.

Cold, warmth, hunger, hedonism and the topography of an entire continent are squeezed into this picture, but from a snow addict's perspective it's a disappointment. The snow is everywhere, but thin. There is only the lightest dusting on a stand of trees in the foreground, and it barely covers the hunters' toes. Some art historians have assumed 1565 was a big snow year and that this was Bruegel's inspiration. If so his painting is a model of restraint, and in fact the meteorological record is unclear. What is clear is that he had a lucrative commission from a Belgian banker to paint all four seasons. Leaving snow out was not an option.

Bruegel would nowadays be called an influencer, and *Hunters in the Snow* started a trend. For the next few decades snow was in in the Low Countries. One of Bruegel's contemporaries, Lucas van Valckenborch, mastered the rare skill of painting it falling, and anyone who likes watching snow fall will find his *Winter* (1586) therapeutic. He paints a scene of 16th-century suburban bustle and then smothers it with so many fat, agglomerated flakes that you can practically feel them on your nose and underfoot. This is the real thing. The effect is powerful enough to give a sense that if you wander off to make a cup of tea the view will have changed dramatically when you come back. But like the snow in Turner's avalanche, it is exceptional. Even van Valckenborch didn't paint snow often, and as the 16th century gave way to the 17th, a depressing thing happened to the output of the Flemish school: snow gave way to ice.

Reality seems to have been intruding. In 2005 Peter Robinson, a geographer at the University of North Carolina, set out to show that art echoes life in the depiction of snow. In a paper for the *Proceedings of the Royal Meteorological Society* he made two connections. The first was between winter severity from 1500 to 1900 and the North Atlantic Oscillation. The NAO, to recap, is the unstable balance between two semi-permanent features of Europe's weather: the Azores high-pressure zone and the Icelandic low. When both are well developed the NAO index is positive. They drag moisture north-east from the mid-Atlantic towards Europe and in winter much of it tends to fall as snow. When the Azores high and the Icelandic low are feeble the index is negative and north-European winter weather tends to be cold and dry, with miserly amounts of snow.

Skiers like a positive NAO. So do farmers. In fact whole sectors of the European economy have a deep interest in what the NAO is going to do next. To make a reliable forecast it helps greatly to know what the oscillation has done in the past, but unfortunately there were no pressure gauges in Iceland or the Azores in the Middle Ages. Despite this it has proved possible to track the NAO from 1500 onwards by measuring some of its effects. These include the depth of snow laid down each year on Greenland's ice cap, measured by studying ice cores; the width of ancient tree rings in Europe, Morocco and the eastern United States; and precipitation in southern Spain, where rain is precious and has been carefully measured for many centuries.

In his 2005 paper Robinson plotted, on one graph, NAO activity as measured by these proxies, and a separate index of winter severity compiled from documentary sources by a team of Dutch meteorologists.* The two lines matched, tracking each other closely for 400 years to the turn of the 20th century, when the study stopped.

This was as Robinson had expected. For his next connection he had to step out of his comfort zone and become an art historian. Could he show that the subject matter of north-european landscape painters reflected the movements of the NAO? Broadly speaking, he could. These artists weren't visual diarists, but they were aware of the changing weather and let it influence their work. The snow that Bruegel painted in *Hunters in the Snow* was probably based on what he'd seen from his Brussels window: a meagre covering from a sky that did not contain the promise of much more. There is little evidence of heavy snow in northern Europe in the winter of 1564–5. It was just extremely cold, including in England, where the Thames froze and Queen Elizabeth I was often seen on the ice.

The following winter was different all over Europe, but especially in the mountains. "Snow was so deep in the eastern part of the Swiss Alps that cattle could not be moved from one stable to another anymore and starved as a consequence," one avalanche

---

* Don't try looking up Winter Severity Index online unless you are also interested in experimental synthesised rock. It turns out to be the name of a band.

study records. Huge snows came again to Switzerland in the winters of 1572–3, 1573–4 and 1575–6. They set off dozens of avalanches that crushed homes and killed their occupants, and they contributed to the Grindelwald Fluctuation, which is not a Robert Ludlum novel but a period of high-volume snow that replenished and extended Europe's glaciers in the late 16th century – the peak of the Little Ice Age. Van Valckenborch caught the snowy mood on behalf of the Flemish school, but by 1630 it had turned again and a father-son team by the name of Grimmer and Grimmer (Jacob and Abel) were painting eponymous scenes of dull grey skies and dull grey ice.

For the rest of the 17th century the NAO is negative, the winters are cold and dry and what snow is available for the walls of Europe's bourgeois homes is mainly in the form of Bruegel knock-offs. Robinson's connection holds, in other words: not much new snow is painted because not much falls. Then, around 1700, the connection falls apart. The NAO turns sharply positive. Winters ease up. The climate pendulum swings back from ice to snow – but artists pay no attention.

For nearly two centuries snow hardly features in European painting. It doesn't return to vogue until Claude Monet gives the signal to his fellow Impressionists in the 1870s, and then it is as if aeons of pent-up snow feelings are vented all at once. The search page of the web gallery of art, which holds 45,000 reproductions of European artworks from 11 centuries, shows how suddenly this happens. Enter the word "snow" in the title field for artists born between 1500 and 1550 (this includes the elder Bruegel, born in 1525) and six paintings are returned. For the next half-century the number is two. For the next 200 years,

1600 to 1800, European artists produce a grand total of three notable paintings with snow in the title. For the first half of the 19th century the number is nine. For the second it is 53.

In fairness to Robinson, changing climate clearly has a good deal to do with this. From 1770 to 1850 the NAO is on such a pronounced upswing that it stays firmly positive even after it peaks around 1860. The winter severity index points even more unambiguously to snow rather than ice in the last quarter of the century, and increasingly detailed weather records confirm that these were snowy times. But there were other reasons for the Impressionists' sudden infatuation with snow. The advent of railways allowed them to head out into the countryside, experience it in the raw even if only for the day, and warm up on the way home. And then there was *le Japonisme*.

Monet's artistic influence on his peers is hard to overstate, and the same is true of Japan's influence on him. Plastered over the walls of his home in Giverny, half way down the Seine between Paris and the sea, were 231 Japanese woodcut prints, many of them depicting snow in a luxuriant, celebratory style that does not seem to have occurred to a single western artist at any point in at least the two previous millennia. Monet didn't even have to own Japanese art to appreciate it. Hundreds of prints introduced France to Japanese ways of seeing the outdoors at the Exposition Universelle, held on the Champ de Mars in 1867. The work of two masters of the form, Utagawa Hiroshige and Katsushika Hokusai, were represented in the exhibition, and their treatment of snow had certain things in common. They showed it falling. They revelled in its effects on natural light, and they showed humans enjoying it.

Hokusai is best known for his *Great Wave off Kanagawa* (with fingers of spray that seem about to pounce on a narrow wooden boat) but in the same series he produced a gentler scene, *Morning after a Snowfall at Koishikawa*, that captures to perfection the simple thrill of waking up to a fresh dump.

It helps that this print captures an exhilarating view of Mount Fuji. Being used to heavy snow thanks to Japan's geography may also mean that its artists are uninhibited about laying it on thick. But the real pleasure in this image is that communicated by a girl on the balcony of a restaurant in the foreground, pointing giddily at the mountain in the distance. The sky is blue. The air is clear. The girl is not hunched against the cold – she's properly dressed for it. She's part of a group eating breakfast *en plein air* for the sheer joy of it. She knows how to live.

Until the opening of Japan to America and the west in the 1850s, most European snow art was, frankly, bleak by comparison. The colours are unfathomably gloomy. The time of day usually feels like late afternoon, with the night drawing in and a sense that people would be better off indoors. In fact I can't help seeing in pre-Impressionist snow paintings, rare as they are, the deep roots of the depressing tradition among western weather forecasters of calling snowy weather bad.

With the arrival of *le Japonisme* the bleak look changed. Pissarro noticed how an old stone house didn't have to cower under snow. It could glow in the reflected light instead. Alfred Sisley let the sun come out over a snow-covered lane outside what is now the Paris suburb of Argenteuil. He realised it would be perfectly natural to paint a family out for a stroll along this lane, just as Hiroshige shows two ladies conversing in the snow rather than enduring it in a famous woodcut made in 1853. Who would have thought? Snow could be fun.

For snow chroniclers this new attitude is good for morale, but also frustrating. Neither the Japanese nor those they inspired could be bothered with putting dates to the snow they painted. There were exceptions, though, and one was Paul Signac's *Snow: Boulevard de Clichy* – a scene of tremendous meteorological energy in which swirling snow is the main event, covering the streets of Paris faster than passers-by can cross them. It is dated 1886, and Signac seems to have been painting real life. 1886 was a big snow year on both sides of the Channel, and we know this because that very winter a tireless snow spotter named Leo Bonacina, one of the most committed to the cause in the whole history of snow record-keeping, noted in his ledger: "Great blizzard."

CHAPTER FIVE:
THE WRONG KIND OF SNOW

"No. It was a different kind of snow"
Terry Worrall, BBC Radio 4, February 11th, 1991

The great British blizzard of 1886 was either an affront or an adornment to an empire in its pomp. No one seemed able to decide which. Autumn had already brought dustings, but the big snow came on January 3rd, a Sunday. A foot fell in seven hours on London. In the Yorkshire Wolds fierce winds off the North Sea pushed the snow into 12-foot drifts. Across the Pennines the Furness railway line to Barrow was blocked for the first time in 30 years. By the Monday, livestock and their keepers were in mortal danger. The papers told a terrible story from the hills of Landrillo in Wales: "Henry Davies, aged 21, while carrying a sheep in his arms, was suddenly blown by the violence of the wind into a deep pool and drowned, in the presence of his father, a shepherd, to secure whose safety he had gone up the mountain, and who could do nothing to rescue him."

In the capital there was no music to accompany the changing of the guard in Buckingham Palace, but *The Times* reported that

Landseer's lions in Trafalgar Square were "enveloped in a coat of dazzling white".

In Kensington Palace Gardens trees buckled and snapped with the sound of rifle shots. Suburban rail services ground to a halt because the telegraph wires between signal boxes were down. Prevented from getting to work, Londoners enjoyed the snow in the royal parks, on Wimbledon Common and on Hampstead Heath. Among them, on the heath, would in all probability have been the family of Leo Claude Wallace Bonacina. He was three at the time. His blizzard of 1886 would have been experienced mainly from a perambulator, but on some level it must have made an impression because the next time snow came to southern England in earnest it set the course of his long life.

Bonacina's father was an Italian merchant who had settled on Haverstock Hill in north London and had grand plans for his three sons. Leo was the oldest, and probably a disappointment. Like his brothers he was educated at private Catholic schools. He attended lectures in many subjects at King's College, London, but his maths wasn't good enough for him to enrol full-time. He never earned a degree. He never established himself in what his father might have thought of as a career, although he did work for 45 years in the library of the Royal Geographical Society. He never married. His true love was snow.

Leo loved the idea of snowbound England, of England as a place of blizzards, and he hoarded every piece of evidence he could find to support it. In 1949 he would write in a letter to the *Journal of Glaciology*: "I am old enough to remember as a child that very severe season when a large meadow at the back of my home in a London suburb was continuously white from 27 November to 21 January." That same winter a family of Bonacina cousins in Devon "spoke of being continuously under snow for several weeks, thus supporting my own experience in Middlesex".

The letter was in reply to one from Eric Hawke, an editor of the *Snow Survey of Great Britain*. Hawke had taken issue with an earlier claim by Bonacina that not just the meadow behind his childhood home but "most of England" was continuously under snow from November 1890 to January 1891. This was not true. There was weather station data to prove it wasn't, but Bonacina preferred to trust his boyhood memories.

As an adult he was short, slight, bald and bespectacled, with a pooterish insistence on being read in academic journals and heard in academic meetings even though he was no academic. For fans of Wes Anderson films, Bonacina was to the RGS as Max Fischer was to Rushmore.

His approach to science was unscientific and he was often ridiculed for it. One 1916 paper that he finagled into the *Quarterly Journal of the Royal Meteorological Society* on a subject outside his comfort zone "was savaged by an audience that had become bored and irritated by his style", according to an otherwise kind obituarist. Amazingly for someone who led both a sociable and a lonely life, he doesn't seem to have let criticism harden his heart. Perhaps he knew his quasi-academic offerings were seen as a charade, and didn't mind, because he also knew he'd found his calling. He was a snow hobbyist, and as a snow hobbyist he was an inspiration. What does this mean? It doesn't mean he was a skier or a Cresta Run hero. There is not even much evidence that he like being *in* snow. He just loved remarking on it, especially when it wouldn't melt. He invented the term "snow survivals" for the patches of snow on chilly British hillsides that sometimes last long into the summer. And ten years before the official start of the *Snow Survey of Great Britain* in 1937, he began publishing regular snow cover reports in the journal *British Rainfall.* He backfilled them to 1875 with the help of Victorian weather data and documentary sources, and bequeathed them to disciples who have gone on updating them to this day. The result is a continuous snow record nearly 150 years old with no equivalent in many countries that have much more snow.

This may seem like a paradox, but only at first sight. Snow fascinates the English in the same way football does, or did until the 2018 World Cup; it is so often such a let-down. That is my experience of English snow, at any rate. In 52 years I've known one English white Christmas, one decent storm in Cumbria and one in Dorset. Also, three episodes worth remembering from 14 years in south-east London. One of them, the March 2018 Beast from the East, dumped substantial snow outside London. None left more than a few inches in the city itself. And that's it. And the iron law of scarcity is that it drives up price and value, and that is the best explanation I can think of for an English attitude to snow – three parts fascination, two parts dread – that you might not find in Troms or Calgary or Magadan.

The updated Bonacina snow record is colour-coded for four degrees of snowiness: "Little" (green), "Average" (orange), "Snowy" (French pink) and "Very Snowy" (blue). In 141 entries, one per year from 1875 to 2016, there are 55 greens, 40 oranges,

38 French pinks and just eight blues. There have been only two blues since 1963. Those came in 1978–9 and 2009–10. Britain is not a snowy country. But snow, like politics, is autobiography. An orange year can be remembered as pure blizzard if you were in the right place at the right time, and a blue year can pass you by if you were distracted or elsewhere.

My favourite entry in the record is for 1981–2. It's pink. It lists December 1981 as the main snowy month of the winter, and it specifies: "12-18th Dec., south west England, 7 in. 20th Dec., northern England, 7 in, 6ft drifts. 6-15th Jan. general snow, 1-2 ft cover."

I remember those December snow days better than yesterday's rain. Much better. I was at boarding school at the end of a long term which the snow miraculously shortened. The whole town went quiet; back to the Middle Ages, when most of it was built. The school carol service was held by candlelight. We wandered around in luminescent darkness, hurled snowballs, and were sent home early.

There are other British snow years that deserve special mention, not always for reasons Bonacina would have understood. For instance:

1620

It has been a long time since any part of the British Isles were properly poleaxed by winter; defeated and suffering grievous loss of life, like Napoleon's army on the retreat from Moscow. In that case the dead were soldiers who could not feel their frostbitten fingers burning as they held them too close to fires in the snow. In the British case the dead were sheep. They were victims of a fortnight-long storm that paralysed Scotland soon after its union with England towards the end of the Little Ice Age.

These were known as the "13 drifty days". They are often dated to 1674 but their most vivid description, by the poet and novelist James Hogg, says they must have occurred in 1620. We should take his word for it. Hogg was Scottish and a shepherd as well as being a writer, and this storm, after all, was mainly Scottish and killed mainly sheep.

"The traditionary stories and pictures of desolation that remain of it are the most dire imaginable," he wrote in the *Gentleman's Magazine* in 1828. "It is said that for 13 days and nights the snow-drift never once abated… and during all that time the sheep never broke their fast." So there was no breakfast for the poor creatures, and no lunch or dinner either. He went on:

> About the fifth or sixth day of the storm the young sheep began to fall into a sleepy and torpid state, and all that were so affected died over night. About the ninth and tenth days, the shepherds began to build two huge semi-circular walls of their dead in order to afford some shelter for the remaining of the living, but they availed but little, for at the same time they were frequently seen tearing at one another's wool with their teeth.

By Hogg's account even sheep that survived the worst of the storm in more sheltered farms died afterwards from hunger and hypothermia. He estimated that nine-tenths of southern Scotland's ovine population perished.

1784

The winter of 1783–4 was the start of a three-year cold patch with two lessons for posterity. The first was that England can occasionally do a passable impression of Canada in midwinter despite the temperate waters that surround it. The second was that when this happens climatologists cannot agree why.

Snow came early in 1783 – in October – and again in 1784. In both years it lingered for months. In both the Thames froze over. 1785 brought the third consecutive year of October snow, which was enough to make reasonable people see a trend. In fact it seems to have been an anomaly.

Benjamin Franklin, who spent the early 1780s in Paris as the American ambassador to France, was among the first to link these very serious winters to a series of eruptions of Laki, an Icelandic volcano with an unusually pronounceable name. The eruptions started in the summer of 1783 and would certainly have disturbed the weather. As their dust and ash spread around the world they would have cast it in shadow. They would have trapped some of the heat already in the troposphere, but reflected more back into space. And they would have provided dust for snowflakes to form around. So it is not a myth that volcanoes can deliver a natural nuclear winter, but whether Laki was solely responsible for the 1784–6 freeze-up is another matter.

Two and a half centuries after Franklin floated the Laki hypothesis, scientists from Columbia University in New York made a discovery. Like Peter Robinson of the University of North Carolina, they had been looking at the history of the North Atlantic Oscillation. Unlike Robinson, who was interested in how the NAO influenced art, the team from Columbia looked at its interaction with the El Niño effect.

El Niño can transform weather across the globe. Californians know this from elementary school, because for them it's local – a periodic warming of the surface waters of the central Pacific. Their coast is on the pacific and that is where the effect is felt first, but everywhere feels it in the end. This stands to reason because the Pacific is the world's biggest water reservoir. When warmer than average it releases more water than usual into the atmosphere. The water is carried east by the prevailing winds across the continental United States, the North Atlantic and Europe, and the storms in which it falls to Earth happen further south than usual.

The questions the team from Columbia wanted to address were how often a strong El Niño coincided with a strong negative NAO, and what the result was when it did. The answers were rarely, and snow. In fact, the team was initially studying the great snows of 2009–10, not 1783–4, and they ended up explaining each by reference to the other. These appear to be the only two winters in recorded meteorological history when an El Niño and a negative NAO of such intensity joined forces.*
Their effects were felt on both sides of the Atlantic, and in 2010 I experienced them in Washington.

Four feet of snow fell on our balcony in two days. That is not enough to break records in North America but it was enough to

* If you are wondering why a negative NAO is associated with snow here, when a positive NAO is associated with snow in other studies, you are probably not alone. In the absence of an El Niño effect, a negative NAO tends to mean cold but dry winters in western Europe and a positive one tends to mean more precipitation, falling as snow in mid-winter. But *with* El Niño, all bets are off.

shut down the federal government and every school from Philadelphia to the mouth of Chesapeake Bay. There was nothing to do at work but write stories about the weather, and nothing to do at home but hurl my two small boys into deep, soft snowdrifts. And there wasn't even a volcanic ash cloud to blot out the sun.

2010 was good, but 1784 had everything. It is on my shortlist of years when the ultimate snow event might actually have happened. Whether it did, we may never know. Extrapolating backwards from recent data is an uncertain science, and documentary sources are scarce, especially from mountainous regions where snow falls hardest.

With the arrival of newspapers, that changed.

1814

In January 1814, as Britain's soldiers manoeuvred to encircle Napoleon in France, their homeland was immobilised by snow. A three-day storm blew in from the Atlantic, hitting the southwest first. On Monday, January 10th it covered Devon's moors and roads to a depth of several feet. Exeter had seen nothing like it in 40 years. Horse-drawn carriages couldn't move. The mail was carried instead by solo riders who could cut a narrower path through the drifts but who risked their lives in doing so. On the 12th a soldier was found dead in the snow near Chudleigh with two shillings in his pocket. Three more soldiers were dug from a nearby drift the following day.

The snow moved north and east and fell on frozen water as well as land. The Thames froze, as did the River Severn and the Solway Firth, which seemed, one witness said, "like a large plain filled with hillocks of snow".

More unfortunates were stranded and froze to death near Burford on the edge of the Cotswolds. One was a post boy, another a farmer found sitting up dead on his horse.

Oxford was completely cut off. The closest any coach came to it before a February thaw was from Banbury to the north. Even that stuck fast two miles from the city, leaving its passengers to struggle on foot through neck-deep drifts to the village of Wolvercote.

What struck *The Times* most was the blizzard's effect on information. The paper's weather news began each day with a list of mail coaches delayed and by how long (mere deaths were buried at the bottom of the story). On January 26th a dispatch appeared that had been sent from Plymouth nearly two weeks earlier. Above it the paper noted that readers would "see that they have earlier intelligence from Germany, Holland and France than they have from their friends and correspondents at home". The writer went on: "This state of affairs is unprecedented in the annals of England."

The empire would reach and pass its zenith before England saw so much snow again.

1947

Here is Bonacina's summary of what may go down as Britain's snowiest winter since the Industrial Revolution:

> Overall grade: Very Snowy. Months of notable falls: DJFM. Outstanding features: Probably worst since 1814… 22nd Jan.–17th Mar., continuous snowcover. 28–29th Jan., south west England, Scilly 7 in. 7th Feb., south Midlands and East Anglia.

25–27th Feb., northern England and Wales, eastern Scotland. 4th Mar., blizzard, Wales, Midlands, northern England, 1 ft; 5 ft accumulated in hills. 12th Mar., Border country.

Footage survives of the 1947 snow and some of it captures respectable-looking depths. In one short film shot on pre-war 8 millimetre stock, Colin Horner of Halifax captures his son, home on leave from the RAF, floundering gamely in drifts as high as drystone walls. But in other films the depth of the snow is less striking than the fuss made of it.

Pathe News sent its own plane on a recce over north London, up to Nottingham and back via Luton. "Coming to Nottingham, familiar landmarks disappear under the white uniform," the narrator says, only they don't. Every shrub, hedgerow and chimneypot is visible from 2,000 feet. "This is Leicester, and here's Leicester's mainline station, but where are the trains? Heading back towards London we find the trains marooned in the deep snow."

The trains are indeed marooned. The question is why, because in fact the snow barely covers the rails. There was enough for skiing on Box Hill in Surrey; enough to warrant the deployment of Polish troops to keep the north–south trunk roads open; enough for Pocklington in Yorkshire to be cut off until German prisoners of war fought their way to it with packs of food. The snow lay longer and in places deeper than in many years, but this was not a winter when nature won. It was a winter when British organisation and ingenuity lost.

The snows of 1947 came less than two years after victory in war. The empire was being dismantled. Rationing was still in force, and now this. What had happened to "can do"? "The equipment for clearing snow and for coping with its effects is primitive and inadequate," *The Times* thundered. There had been experiments with flame throwers which had failed, and with jet engines "which evidently merit further investigation for clearing fresh, deep snow drifts". This turns out to be true. There is nothing quite like hot gas and several tons of thrust to shift snow fast, which is why jets mounted on lorries are used to clear taxi-ways at airports across Russia. But then, reaching for something to say, *The Times'* editorialist conflated weather and climate and forgot that London is not Moscow:

> It is a national foible to refuse to admit that the British winter is anything but mild... It would seem an essential precaution when the margins of the national economy are so narrow, that before next winter the subject of snow should be reviewed and the necessary equipment acquired. The respective merits of the rotary plough and the jet engine must be ascertained. A sufficient number of suitable machines must be obtained and disposed at strategic points on the railway and roads to ensure that coal and other supplies can be moved to the right places without delay in all conditions.

Memory plays tricks, just as editorialists write nonsense. Britain's weather conditions had been unsnowy for three of the four previous winters and would be unsnowy for two of the next three. With the margins of the national economy so narrow it made no sense at all to spend heavily on rotary ploughs and jet-powered snow removers. The typical British winter was not "anything but mild". It was mild. It is mild, and getting milder, which is why anything else attracts so much attention.

1963

There are people still alive who remember the winter of 1963 as their personal mini ice age. Bob Dylan was living in a flat in London at the time and had to burn furniture to stay warm. It was cold throughout Europe; the only winter in the 20th century in which Lakes Constance and Zurich froze.

By November, snow blanketed south-western England. It was topped up in February by two extra feet from Devon to the north-east. Inevitably, this weather was called "bad". But how bad was it? The truth is it was a severe winter by British standards but not by those of any other country on the same or a higher latitude. It stands out in the record for two reasons: it was 3.7°C colder than usual, and this meant that when snow fell it stayed. Both these features were the result of a strong negative NAO that made its presence felt further west than usual. In 2007 researchers from the University of Southampton compared that negative NAO with those of 1955–6, 1968–9 and 1984–5 and found that all three delivered colder, snowier weather than 1962–3; they just delivered it further east.

"What makes the winter of 1962–3 extraordinary [for Britain] is the westerly location of the cold temperature anomaly," Dr Joël Hirschi of Southampton's National Oceanography Centre wrote. This showed how "a shift in temperature by a few degrees C can transform the traditional green British winter into a snowy (white) one".

So the significance of 1963 was this: for once in their lives the British got a taste of other Europeans' winter. London was Vienna. Norwich was Kiel. It was the exception that proved the rule that in Britain snow is even more exotic than sun (and no one has a clue how to handle it).

1991

One of the great team contributions to climate science is the British Met Office's Hadley Centre Central England Temperature series, or CET for short. It is the longest continuous temperature record for anywhere in the world. It has been updated daily from 1772 to the present and in its rawest form consists of about 7,600 rows of numbers each 12 figures long. Overall, it shows a long-term mean temperature rise of about a degree, most of it since the mid-20th century, which is something we are all going to have to work out how to live with or reverse. But buried in the numbers are clues to thousands of less insidious weather stories, some of them specifically and famously involving snow. For instance, the days from February 3rd to 14th, 1991, are represented by the following numbers: -9, -12, -8, -15, -47, -36, -38, -28, -5, -20, -1 and -9. Each number shows the mean temperature for one day in tenths of a degree, so these 12 days were not Arctic, but they were cold. The temperature almost never rose above freezing, and from the 5th to the 10th it snowed over most of the country.

Unusually for Britain – but predictably given the low temperatures – this snow was light and powdery, not wet and sticky. London got five inches. Many areas received a foot, and that was enough to wreak a special sort of havoc.

British Rail, then still state-run, had recently taken delivery of 7,000 new electric propulsion units, most of them passenger carriages incorporating motors rather than old-fashioned locomotives. The snow caused 35 per cent of them to burn out. In addition, points froze and sliding doors stuck fast. On February 17th, looking back on the storm, *The Sunday Times* reported that "two of London's biggest terminal stations, Euston and Waterloo, were cut off and three quarters of the rolling stock for 50 miles around the capital lay marooned". On the 11th, a Monday and a hellish one for commuters, Terry Worrall, British Rail's head of operations, had bravely consented to be interviewed by James Naughtie on the BBC's *Today* programme. Asked what the devil was going on, he said: "We are having particular problems with the type of snow, which is rare in the UK."

NAUGHTIE: Oh, I see, it was the wrong kind of snow.
WORRALL: No, it was a different kind of snow.

Worrall was quick to correct the record, but not quick enough. Naughtie's summary resonated up and down the land. That afternoon "The Wrong Kind of Snow" was the headline above the *London Evening Standard*'s splash. The die was cast. The phrase entered the vernacular as swiftly and easily as a certain sort of Etonian enters parliament. It has been the dominant exemplar of pathetic British excuses ever since, and is the title of a book about them. It also probably hastened the privatisation of British Rail, even though the amiable and honest Mr Worrall never actually said it.

What had happened, he said when given a chance to explain

himself in more detail, was that strong cross-winds had blown this un-British fall of perfect powder straight into the new rolling stock's side-facing air intakes. They were there to cool the motors, and designed to channel away rain and any snow that responded predictably to gravity. This was not that sort of snow. It floated around until it made contact with hot electrical components, whereupon it melted and shorted them. It did not help that the intakes were not fitted with filters; nor that the new motors were mounted low, below floor level, unlike those of continental rolling stock of the same generation, which could be mounted higher without unduly obstructing passengers because it was wider.

"You can't know that there are going to be problems with a certain type of snow," Roger Freeman, the railways minister, told the *Guardian*, but of course he was wrong. You can know. All he had to do was look at the weather forecast and then at the Nakaya snow diagram, which would have told him that at the predicted temperature and humidity levels the snowflakes that fell were going to be mainly small and streamlined. They were going to be plates and needles, with some smaller dendrites but few big ones. But there is no evidence Freeman knew any more about snow than Worrall. *The Times* devoted a sarcastic editorial to their complacency: "Because it is never going to snow again, British Rail seems to tell itself, why worry about the design of all the points, doors, motors and signals which add up to a modern railway system?"

Why worry? It would have been a different story, eliciting more sympathy for British Rail, if the snowfall of February 5th–10th had been truly daunting to shift, but it wasn't. "It was different snow, but not an awful lot different," the Met Office said.

The search for the most abject surrender to snow is over. The search for the ultimate snow event continues.

CHAPTER SIX: RECORD BREAKING

"Outside a blizzard was raging; the wind was howling, the shutters shaking and rattling; everything seemed to her like a threat or a mournful omen."

Alexander Pushkin, *The Blizzard*

Big snow is so exciting that it's useful to have a benchmark to help you know when you're dealing with the real thing. In 1938 Charles Franklin Brooks, founder of the American Meteorological Society, provided one.

"A question has been raised as to how much snow can fall in one day," Brooks wrote in the society's bulletin, which he also edited. The answer is not straightforward, but he made it look that way. Starting not from first principles but from observation, he took the world rainfall record for a single day, which at that time was 46.99 inches in a severe tropical cyclone that hit Baguio in the Philippines on July 14–15th, 1911. He rounded 46.99 up to 47, divided it by five and multiplied it by ten.

The division was necessary because the rain fell on Baguio in air at an average temperature of 24°C, which can hold five times as much moisture as freezing air, which is necessary for snow.

The multiplication was necessary because, as a rule of thumb, allowing for wide variations depending on a given snowstorm's water content, snow is ten times less dense than rain. You get ten inches for every inch of snow-water equivalent. The provisional result was an estimate of 90 inches or 229 centimetres as the maximum snowfall for a day. But Brooks, a Harvard man who had been raised in Minnesota, knew something about snow. It settles. As it accumulates, new snow weighs down on what lies beneath. Or, as he put it:

> The weight of the snow would be so great that the density of the entire layer could not be so little as one in ten. Therefore the snow could hardly average more than two thirds or three quarters of the normal density. Taking two thirds as a conservative value, the maximum possible depth of snowfall in one day might be considered to be approximately six feet.

So Brooks came up with a nice round number just big enough to bury most of humanity standing up, leaving only those of exceptional physical stature with any sort of view. Still, his benchmark turns out to be a good one. Subsequent snowstorms have shown that six feet or roughly two metres in a day may not break records but it will come close, and it will cause trouble.

For instance, on December 18th, 1991, a strong mistral blew in across France from the Atlantic. The wind was driven like steel through a rolling mill by huge rotating pressure zones – the usual Azores high and a deep low over the Hebrides. A second low over the Gulf of Genoa added Mediterranean moisture.

France was gearing up for the 1992 Winter Olympics in

Albertville. Thousands of tonnes of high explosives had been used to widen the old road up the Isere Valley to the ski resorts of the Tarentaise, but with two months to go the new road was not yet fully open. From Albertville into the mountains it was one lane each way.

Storms swept the Savoie for a week. They reached a crescendo on the 21st, a Saturday. The moisture fell as snow and the result was madness. From ten in the morning nothing moved on Route Nationale 90. It felt as if everyone was there; Parisians heading to the mountains for Christmas, builders stuck in vans when they were supposed to be working overtime for the Olympics, and bus-loads of Brits.

Darkness fell. The snow fell harder. Still nothing moved. The only vehicles not lit up red by brake lights were those whose drivers had switched off their engines. I was in a bus, on a jump seat at the front. The other passengers had each paid thousands to be stuck in this jam. They were my responsibility and they were getting antsy because by this time they had expected to be installed in warm chalets in Courchevel, the most pampering of all French ski resorts.

The snow was unrelenting and magnificent: coin-sized agglomerations of flakes landing like cargo palettes on parachutes. They were messengers from the full-force blizzard we knew was raging higher up, but it was hard to welcome them as they deserved because my passengers were more concerned about comfort stops and wine selections.

For long periods the bus's big solo windscreen wiper was the only thing that moved. It was hypnotic. When at last the bottleneck cleared we could start to climb, and the higher we climbed the harder the wiper had to work. It was driven by an electric motor at the base of the windscreen that made a rhythmic buzzing sound. I could have chartered a newer bus, and should have. After eight hours on the road, around midnight, it nosed into the centre of the resort and sighed to a halt in a snowdrift. There were chains on its drive wheels but it could go no further. We were stuck. The electric motor was on fire and the bus was filling with smoke.

The snow was so deep and soft that only one vehicle in the whole town could still move. It was a taxi, but not just any taxi. It was a black Toyota Land Cruiser with non-slip differentials and partially deflated winter tyres that could grip the snow at any angle. It was driven by a man possessed. I paid him to distribute my charges to their lodgings and was told by one of them the next morning that he would be suing me for failing to switch on his heating. The engineer I'd booked to fix it had been stuck in traffic.

The snow continued until Christmas Day. The French meteorological service categorised the *événement* of 18th–25th December, 1991 as a two-metre storm. It's etched in my memory for the ferocity of the snow and the rage of my litigious guest, yet in the record books it barely features because it was spread out over a week.

Snow is more likely to make headlines when it comes all at once, as it did on October 12th, 2014. A few days earlier a powerful tropical cyclone from the Bay of Bengal made landfall near the Indian city of Vishakhapatnam and rolled north towards the Himalayas. There it trapped hundreds of trekkers in a blizzard – many would not survive. Paul Sherridan, a police officer from South Yorkshire, was one of those who made it out alive.

"The first I was aware of some sort of weather event was on the morning of the day before," he said when I reached him on the phone a few years later. He was trekking with a guide and two other westerners, a Dutch couple, around the Annapurna circuit in north-central Nepal. They were billeted in a guest house in the tiny village of Thorung Phedi, and the next stage of the trek was the highest: a climb to the Thorung La Pass at 5,099 metres.

There was a weak cell phone signal at the best of times in Thorung Phedi. Their guide, like most, had little contact with the outside world, but Sherridan overhead a briefing given to a larger group of trekkers. It included "the possibility of a bit of bad weather".

"The next thing I knew of the event was when it was snowing," Paul said. By then it was about 4pm on the 13th. Thorung Phedi is nearly as high as the pass and bitterly cold at night. As the sun went down the snow started to settle. By 8pm there was a good covering, although still only a few inches deep.

Trekkers usually start the ascent to the pass well before dawn. Most of those with Paul at Thorung Phedi were young Israelis celebrating completion of three years' military service. This did not mean they were well equipped, and ordinarily this would not have mattered. October is peak trekking month in Nepal because the skies tend to be clear, temperatures tolerable and views spectacular. Many of the Israelis had only training shoes to walk in and clothing adequate for autumn but not winter. As a result their decision on waking on the 14th was more difficult than expected: to head for the pass as planned even though it was snowing; to sit it out at Thorung Phedi; or to retreat to a more comfortable altitude.

Paul had good boots and a weatherproof jacket. He was an experienced hiker and in good physical shape for 48. He headed out into the snow.

"Hundreds of people set off at the same time, all in different states of dress," he said. "Most were walking but some had hired mules or horses. The guides had wrapped plastic bags round their plimsoles to keep the snow out. People were using bin liners as anoraks."

The first hour and a half of the climb was by torchlight. Despite the snow, dawn was more grey than white. It revealed no horizon or boundary between snow and sky; only a long line of figures in varying degrees of distress.

"By eight in the morning it was clear that some of the guides were disorientated because they weren't used to snow. People were collapsing. Horses were running around." At the pass there was a low stone shelter but it was full by the time Paul reached it. Ice formed on the faces of those left outside wondering what to do next. "People's eyeballs were freezing because they had no goggles. Someone suggested the best way down would be to carry on over the pass, so I joined what seemed at first like an orderly queue, but it quickly became apparent that if I stayed in the queue I was going to freeze to death."

The snow intensified. "At first I could see thirty people, then 30. Then I could see only my hand if I touched my nose."

It was an ambush. A year later a team of scientists from Utah, where snow is more valuable than uranium, pieced together what had happened. They found that of all the tropical cyclones generated in the Bay of Bengal since 1979 only one has survived as a storm all the way to the Himalayas, and this was it. The US Navy's Joint Typhoon Warning Center tracked it for 1,500 miles from the Andaman Islands to Nepal, but few there thought it would do any damage.

"I thought it would be a little rain, some snow, some clouds," the director of a local trekking company told the *New York Times*.

There were reasons not to worry overmuch. Cyclones weaken the moment they make landfall because warm waters are their main energy source. This one, named Hudhud after an exotic crested bird, was no exception. It was losing strength as it arced across north-east India on October 12th and 13th, and the skies over Kathmandu were clear. What happened next was "a classic tropical-extra-tropical interaction", the Utah experts wrote. The tropical component was the cyclone, spinning slowly in an anti-clockwise direction. The extra-tropical component was an unusual but not unprecedented southward dip of the jet stream. The cyclone brought moisture. The "upper level trough" brought cold air from the north-west. They collided over the tenth highest mountain in the world, and there Hudhud stopped and released all that remained of its wet cargo in one extraordinary storm.

Two processes ensured that every last ounce of water vapour was wrung out of the sky and converted into snow. One was orographic uplift, the irresistible rise of an air mass hitting high ground. The other was "tropographic ascent" – warmer air clambering over cool like a great invisible duvet, as it does whenever snow falls without the help of mountains.

The completeness of the event is illustrated by weather maps for October 12th, 13th and 14th showing Hudhud's tail as a smudge advancing on the Himalayas, and the upper-level trough sweeping across them. On the 14th the smudge covers the entire range, centred on Annapurna. On the 15th it abruptly disappears.

To be directly under this collision was terrifying. Paul said:

> It's a bit like being under water and not knowing whether to swim up or down. Then add winds so strong they'll knock you off your feet. Then cold that will freeze your eyeballs, and sound so loud it's like standing next to a plane on a runway. It's just awful.
>
> The snow quickly went from below your boots to above your boots to half-way up your shins and sucking you in. It was wet and heavy, more like sand than snow. At the worst point it was so deep it was difficult to move. I was breaking trail and because the way wasn't clear I'd slide off the trail and up to my waist, and that was at 17,000 feet.
>
> My heart was beating so hard that it seemed I couldn't breathe. I was just fixed to the spot and expecting to pass out. That was the only time I thought I was going to die. The only way I can describe it is if you were told to hold your breath under water beyond the point where you could hold it any longer, and then you open your mouth to take a breath and you can't.

Dozens of people were following Paul; possibly more than 100. He was credited afterwards with helping to save their lives, but many were lost. Forty-seven trekkers and guides died in the Annapurna storm. An unknown number turned round on the descent from the pass and were not seen again until their bodies were dug from snow banks 30 feet high. Long before that, Paul knew what had happened: "As I was coming down I could hear the avalanches that were killing people behind me."

The storm didn't break records for snow depth, partly because its zenith was brief (Paul estimates it was at its most intense for only 15 to 20 minutes on the morning of the 14th); and partly because its water content was so high. This was a slush-puppy blizzard at the opposite end of the snow-water spectrum from a powder dump. The snow started to sinter and compact almost as soon as it landed. But as a frozen cyclone it was rare and must have been close to record-breaking for the speed at which snow fell.

On October 15th a helicopter search party found footprints in drifts six and a half feet deep. Conservatively, that might indicate an average depth of half as much near the top of the pass, which would tally with Paul's recollection of sliding into waist-deep snow when stumbling off the trail. Some of it had fallen overnight but most had fallen in a single morning. Three feet or a metre in eight hours is fast snow, and may have been a harbinger of more to come. Tropical cyclones from the Bay of Bengal are becoming more intense with rising sea surface temperatures, and more northerly. Meanwhile the jet stream in this part of Asia has shifted south by an average of 200 miles since 1948. More collisions are likely, and so is cyclonic snow.

Paul, when I spoke to him, was still curious to know why he had found it so hard to breathe in the worst moments of the Annapurna storm. He had asked a doctor on his return to England and got no satisfactory answer. I couldn't offer one either, apart from the altitude – although that would have affected him more or less equally throughout the day. But his description recalled one presented by an American geographer, Douglas Powell, at the US Western Snow Conference in 2006.

The Western Snow Conference is principally a gathering of hydrologists studying snow because the western US depends on it for irrigation. But occasionally it accepts lighter fare from people like Powell. In his twenties, he had been a snow surveyor for the state of California in the southern Sierras. On February 24th, 1969, he found himself in the middle of a sustained and ferocious snowstorm in the upper Kern River basin that would nowadays probably be attributed to an atmospheric river. To Powell it was simply "the maximum-intensity snowstorm in my over 30 years' experience of field snow surveying in the Sierra Nevada, Afghanistan and Chile". Determined to experience it full force, he put on skis and headed out from a surveyors' cabin at 10,000 feet into a clearing surrounded by trees. It is marked on maps as the Big Whitney Meadow. Snow had been piling up at three inches an hour for a day and a half. It showed no sign of slackening but was being driven horizontally by a 50–-60 mile-an-hour wind.

"Visibility was zero. Breathing was nearly impossible [and] survival demanded an immediate retreat into the shelter of the lodgepole pines, with gratitude for still being alive," Powell told the conference. "What living creature could survive out in that wind and snow? While collecting my senses back in the trees, above the howling of the wind, a coyote, probably within 100 feet of me, let loose a long, reverberating call… Spontaneously I replied with a long, reverberating yell, and for the next 15 minutes we serenaded each other. I do not believe the coyote took me for another coyote, but maybe it sensed, as I did, we were kindred spirits in this world of white."

Agony and ecstasy are not far apart here. If all the world's snowy places had enough snug surveyors' cabins to go round, snow might have a better reputation. But they don't, and snow can simply smother the unwary. In the upper Kern basin, Powell, like Paul Sherridan on the Thorung La Pass, felt unable to breathe simply standing in fast-falling snow. In British Columbia and the north-western US, there is an acronym for fatalities from essentially drowning in snow. They're NARSIDs – non-avalanche-related snow immersion deaths – two-thirds of which occur in tree wells: skiers fall head first into deep snow holes near tree trunks or under low-hanging branches, and suffocate. The holes are formed by localised melting but they offer little in the way of air. Ninety per cent of those who fall into them need help getting out, and if it doesn't come quickly, asphyxiation does.

In Iran, in 1972, snow drowned entire villages. The worst blizzard in recorded history in terms of loss of life laid siege to most of the country for six days. It came from the north-west, charged with moisture from the faraway Atlantic but also from the Black Sea. Snow fell without interruption from February 3rd–8th, from the Turkish border to the high deserts hundreds of miles south of Tehran. Reports from the ground were scarce, especially in English, and as a result a good deal of nonsense has been written about the Iran blizzard since. It features on most lists of historic snowstorms, but many of those who compile them assume it came as a bolt from the blue to a Middle Eastern land unused to snow.

In reality snow is an integral part of Persian life, art and geography. Every winter it blankets the mighty Alborz Mountains north of Tehran. Every summer, snowmelt waters the whole country. Even the arid-looking mountains of Kerman province in the south-east are as used to snow as those of New Mexico, which they resemble. But in Iran, being used to snow is not the same as being prepared for it. In rural uplands, farmers and their families would traditionally sit out a snowstorm "hunkered down in their houses, kept warm by their livestock in the same building," the chairman of the London-based Iran Society told me. This did not pass muster as a survival tactic in 1972.

By February 10th, 6,000 people were estimated to be missing. The final death toll was put at 4,000 but may have been higher. Local newspapers said 200 villages were destroyed including two in central Iran, where no trace of the dead was ever found. The Associated Press reported on a search party sent to the Turkish border zone when the weather broke on the 9th to look for survivors. At a village called Sheklab, population 100, they found none, and only 18 bodies.

After one clear day the snow returned. The headline figure for the depth dumped on Iran in a total of seven days was 26 feet or nearly eight metres. If true, that would surpass the western record for a single storm by 75 per cent. In reality the figure was probably inflated by drifting and avalanching, and may have included snow already on the ground from a series of smaller storms the previous month. The record that the February blizzard definitely holds, unfortunately, is for the deadliest in the annals of snow. The search for other sorts of records gravitates to countries that set more store by them.

Nowadays the most authoritative extreme snow data in the US is gathered in one place by the National Oceanic and Atmospheric Administration and arranged by state. For each state, all-time snowfall records are given for one, two and three days, along with the measuring stations where they were set. Some fantastically rugged place names feature. Who could resist counting snowflakes at Wolf Creek Pass or Last Chance, Colorado? All four great American snow factories are in the mix: the Rockies, the Cascades, the Californian Sierras and the Chugach of south-central Alaska. For the record the current official US record holders are:

For one 24-hour period: Silver Lake, Colorado, on April 15th, 1921: 76 inches (193 centimetres)

For one 48-hour period: Thompson Pass, Alaska, December 30th, 1955: 120 inches (304 centimetres)

For one 72-hour period: Thompson Pass, Alaska, December 30th, 1955: 147 inches (373 centimetres)

The trouble with these numbers is that they skate over

history, and great storms get overlooked as a result. For most of the first half of the 20th century, the US record for one 24-hour period was thought to be 63 inches (160 centimetres), set in Georgetown, Colorado, on December 4th, 1913. Georgetown is now a way-station on the Interstate 70 freeway as it hauls traffic into the Rockies from Denver. In 1913 it was a gateway to the Wild West that the snow slammed shut for several weeks. Sixty-three inches is five foot three, well within CF Brooks' theoretical maximum, so as president of the American Meteorological Society he could sit back and wait for nature to prove him wrong. In the end it was a mixture of nature and diligent scholarship that did the job: in 1953 a hydrologist with the US Weather Service in Washington, J L H Paulus, revisited official Colorado snowfall measurements for a storm that lingered over the continental divide for four days in April 1921. It was a beauty, rolling in from the Midwest and piling up against the front range of the Rockies in a billowing mile-high berm of frozen vapour.

In Denver some of this storm fell as rain but at 10,000 feet it fell as exceptionally light snow. Three miles east of the divide, at the Silver Lake measuring station above Boulder, it built up steadily at a rate of more than three inches an hour for 27 and a half hours. The snow then slackened, and resumed. The total recorded depth for the storm was 100 inches (254 centimetres). Seventy-six inches (193 centimetres) or six foot four inches fell in its first 24 hours; 95 (241 centimetres) in its first 48; and 98 (248 centimetres) in its first 72. High winds caused drifting, especially on the second full day (April 15th). But there was no evidence, Paulus wrote, "to indicate that the Silver Lake observer used less care in obtaining a representative snow depth than did the observers who measured previous record snowfalls".

The kicker was the snow's lightness. At Silver Lake the water equivalent of what fell in those first 27 and a half hours was 5.6 inches, giving a density of 0.06 and an amazing 15.5 inches of snow for every inch of snow-water equivalent. Brooks, remember, had based his six-foot-per-day maximum on a 10:1 snow water ratio. So the great man was not technically in error. He had just not bargained for the very finest Colorado powder. After due consideration, Paulus wrote in the *Monthly Weather Review*, "the Silver Lake snowfall is being accepted as providing the highest known rates in the United States for durations to four days".

That was in 1953. Silver Lake has since been supplanted in the two-, three- and four-day stakes but retains the US record for a single day.

The Alps have not been able to match it. One of their closest contenders was a storm at Bessans in 1959, an hour's drive from Turin on the French side of the Italian border. It came late in the season and hardly touched any of the neighbouring valleys, but along the lonely mountain road linking Val d'Isère and Modane, which only opens in summer, 172 centimetres fell in 19 hours. According to a paper in the *Revue de Géographie Alpine*, the snow had a strange, salt-like consistency that meant it rolled down steeper slopes in a series of mini-avalanches rather than building up to any big ones. Like the Himalayan snow of October 2014, it melted fast. Such are the fleeting joys of spring in the Haute Maurienne.

Japan's snows pile up higher and stay longer. Every year in

mid-March surveyors in the Hida Mountains west of Tokyo send a bulldozer out over the snowpack with a GPS route-finder. The purpose is to mark out a track directly over the main east–west mountain road, which at this time of year can be buried under more than ten vertical metres or 33 feet of snow, enough to bury a five-storey apartment building or two double-decker buses on top of each other.

More bulldozers then clear the road to a depth of two metres. Then rotary ploughs take over. Between them they cut a neat, straight-sided canyon, the Yuki-no-Otani, which attracts tourists until mid-June and proves an important point. However much snow falls where its depth is measured regularly, more snow falls in higher, wilder places where it isn't.

For instance, the snowiest official site in Japan, singled out by Professor Jim Steenburgh as even snowier than Utah's Wasatch Front, is Sukayu Onsen, a hot spring and spa in the northern tip of Honshu. Average seasonal snow depths here are prodigious. The Sea of Japan, offering up limitless water vapour to Siberian winds in the depths of winter when a mere lake would be frozen over, is the snow god that keeps giving. The Japanese Meteorological Agency's record depth for lying snow here is 523 centimetres. And yet it's clear that this depth is routinely exceeded at the site of the Yuki-no-Otani. The canyon walls there are up to 65 feet or 20 metres high. Some of this height is accounted for by snow blown up onto the sides of the canyon by the rotary ploughs. But this is necessarily less than half the total height because the ploughs only remove the last six feet of snow. The rest is pushed along the slot by bulldozers and then to one side where the snow is no longer at its deepest. So there may be further lessons to be drawn from Yuki-no-Otani: official records are good to have but are not necessarily *actual* records. Unofficial ones need not be dismissed just because they are unofficial, and in that spirit I pass on some astonishing Japanese snow numbers provided initially to Christopher Burt, an American extreme-weather historian, by Yusuke Uemura, a Japanese snow enthusiast. All are unofficial. All are measurements taken by Japanese railway staff at stations in the alpine north and west of Honshu. All are for cumulative snowfall in the very severe winter of 1944–5. Here goes:

3,555 cm (1,480 inches) at Oshirokawa Station, Uonuma prefecture

3,280 centimetres (1,291 inches) at Echiyo-yuzawa Station, Yuzawa prefecture

3,126 centimetres (1,231 inches) at Ihirirose Station, Uonuma prefecture

3,090 centimetres (1,216 inches) at Sekiyama Station, Myoko prefecture

3,010 centimetres (1,185 inches) at Tsuchitaru Station, Yuzawa prefecture

The claim in each case is of more than 30 metres of snow – about three times the average snowfall claimed by the snowiest place in Europe and substantially more than the commonly accepted world record snowfall for a single season of 1,140 inches at Mount Baker in the Washington Cascades in 1999.

The numbers may be unofficial but they are appealingly precise. When I got in touch with Yusuke Uemura to ask how they were measured, he replied that measurements by Japan Railways were conducted using a snow scale (a vertical calibrated marker) for snow depths, and a snow board for snowfall. Snow boards are placed out of the wind and cleared off after each measurement for more accurate cumulative totals. The measurements, Uemura said, were conducted twice a day, at 8am and 4pm until 1987, and at 8.30am and 5pm from 1988.

In a separate submission to Burt in 2014, citing the Japanese Meteorological Agency, Uemura stated that the world record snowfall for a single day occurred on the upper slopes of Mount Ibuki, a Ben Nevis-sized lump not far from Kyoto, on St Valentine's Day 1927. The depth: 230 centimetres or 90.6 inches. That far exceeds Silver Lake's 76 inches, but if it was the world record as of 2014 it did not last. The following year it was beaten at an unexpected location not in Japan, to which Japan nonetheless offers a clue.

The clue has two parts: latitude and proximity to water. It is interesting to stand at a map of the world and trace a line west from Sukayu Onsen on the same parallel. Its latitude is about 40 degrees north. Moving west you meet the Asian landmass half way down the North Korean coast. You cross the Caspian a short distance north of Baku, miss the Black Sea altogether and make landfall in Italy roughly on a level with Rome. If you were a gust of wind you would expect to pick up water vapour from the Adriatic. If you were a water molecule you would have to be braced, in winter, for blasts of cold from the north-east, and for orographic uplift and brusque attachment to a snow crystal as you hit the Apennines. They are not very high, but nor are the mountains around Sukayu Onsen. In simple geographical terms the two places have a lot in common, and it turns out they compete for snow records too.

On March 10th, 2015, the small town of Capracotta, in a fold of rough pastureland on the Apennines' eastern flank, was visited by an uncommonly heavy storm. It was "a spectacle that took our breath away," the mayor said. Two and a half metres of snow fell in 18 hours. That is 100 inches – 24 more than fell on Silver Lake in a whole day.

Capracotta is used to snow in winter because it takes the full force of every Balkan nor'easter broadside on, which is why every household has a snow shovel, and why the hills above it are dotted with wind turbines. Despite this there were doubts at first that somewhere 300 miles south of the Alps could have been buried so deep so quickly. But the photographs taken by locals that evening, of snow banks filling ancient streets and rising to first-floor bedroom windows, do not lie. Soon afterwards *The Guinness Book of Records* certified Capracotta as the all-time world record holder for one-day snowfall. Silver Lake was demoted to second place after 65 years at the top, and there the quest for the ultimate snow event might have ended, at least for now. But that would be boring, and unscientific.

So far the main conclusion to be drawn from this quest is that so-called snow records are guides only. They are unreliable because no snow surveying system can be everywhere at once, and even those that are ostensibly well organised can ignore big snow events for no good reason.

This brings us back to Douglas Powell, the Californian snow

surveyor who communed with a coyote for an unforgettable 15 minutes in Big Whitney Meadow in 1969. He did not formally claim any records for that storm when recounting his experience of it half a century later. Instead he called it "world class" and noted that it "exceeded any amount I remembered being recorded anywhere else". But it clearly did break a record, not for one day but for two. Here are its parameters, in Powell's words:

> It began to snow precisely at 6pm, just as we entered the cabin. Unlike many storms, the rate of snowfall was heavy right at the beginning. For the next 48 hours the rate deviated very little from 7.6 centimetres (3 inches) an hour… It stopped snowing suddenly at 6pm on the 24th, precisely 48 hours after it began. With the snow sampler and tape, we measured the newly fallen snow at several unobstructed areas just outside the cabin. The average accumulation in 48 hours was 183 centimetres (6 feet) each 24 hours, 366 centimetres (12 feet) total in 48 hours.

This did not beat Silver Lake for one day, but it handsomely beats Thompson Pass, Alaska, the official NOAA two-day record holder with 120.6 inches. And yet the NOAA has ignored it.

Powell was a decorated war veteran who after his stint as a snow surveyor became a popular geography professor at the University of California at Berkeley. His field trips were always over subscribed and the quality of his research was never questioned. He did not brag about his measurements in 1969 but he did ask: "Who is better qualified to measure record snowfalls than veteran snow surveyors?"

Who indeed? What else has the NOAA ignored? The historical and geographical record is littered with evidence left by great storms never measured. They came quietly, but when the snow they left behind began to move, it roared.

CHAPTER SEVEN: WINTER OF TERROR

"What bothered me most was the sucking sensation, like being caught in a breaking wave. It was pulling me under. 'Oh Christ,' I thought. 'I'm in the white. I'm in the cloud... I'm in the shit.'"

Jim Sweeney, avalanche safety officer, Valdez, Alaska, quoted in *The Times*, February 1st, 2012

At ten to one in the morning on February 12th, 1951, the Swiss village of Airolo at the foot of the St. Gothard Pass was woken by a sound like thunder. Snow had been falling for 36 hours. Above the village a metre of fresh powder had settled on a base of a metre and a half. Two weeks earlier the town of Andermatt on the north side of the pass had been overwhelmed by six avalanches in an hour, and Airolo was braced for calamity. Police had ordered a partial evacuation and army units were on standby. Even so, those who stayed had little time to save themselves.

Detached from the mountains as if by an unseen hand, a curtain of snow tumbled towards the village from the west. A concrete barrier built to protect it in the 1920s might as well have not been there. The snow surged over it in a front 200 metres wide. It advanced into the village, buried the church to half way up its steeple and kept going.

In the early seconds the snow had accelerated to several times the speed of an express train; hence the thunder. Then it slowed, moving for 12 minutes in all before slithering to a halt like a many-headed monster. By 1.15 the school and a dozen houses were destroyed. Fifteen villagers were buried. Five were found alive that night, but it took five days to dig out the bodies of the other ten. By that time avalanches had claimed 91 Swiss lives since the start of the year. The total by the end of the winter would be 99. In Austria it was 135. For the Alps as a whole it was 256, a number that played on the consciences of engineers and architects who thought they understood the power of snow quite well. As it turns out they had much to learn from what became known as the Winter of Terror.

1951 was probably the snowiest year in the central Alps since 1916. In that year Austrian and Italian troops had perfected the weaponisation of avalanches by setting them off with howitzers. They killed, by one estimate, 3,000 of each other in a single 48-hour period in the Tyrol, and about ten times as many over the course of the whole war. "I have seen the corpses," Walter Schmidkunz wrote in *Battle over the Glaciers*. "It is a pitiful way to die."

The snow depths of 1951 were not exceeded until 1999. The terrible confirmation came at tea time on February 23rd. Over the previous month five metres of snow had fallen on the slopes above Galtur in western Austria. That was a record; enough to put the whole region on maximum avalanche alert and cut the village off from the rest of the country. But it was not enough to account for the avalanche that destroyed much of Galtur in two minutes that afternoon. Its scale and violence were completely unexpected. Thirty-one people died, more than in any single avalanche in Europe in 1951 or since. In the months that followed scientists pieced together what happened but could still scarcely explain it.

They put the mass of snow in the starting zone high above Galtur at 170,000 tonnes. They doubled that on the basis that it would have picked up extra snow on its way down. They estimated that it hit the village as a wall up to 100 metres high.

The damage was done not by a solid mass of snow – that stopped well short of the village – but by flying chunks of it, and a cloud of fine, dry powder that reached a top speed of 186 miles an hour. A computer animation based on this sort of data makes the impact look pure Hollywood, but it's the only way to explain what survivors saw: buildings in the avalanche's crosshairs that seemed to explode or were simply ripped from their foundations; balconies torn apart like matchwood; flying snow so thick that breathing was impossible; and when it had settled, a pile up to 20 metres deep where once there had been a car park.

As far as anyone could tell, this is what happened. The irony is that the Galtur story could only be told and tested with the help of a Swiss avalanche research centre whose mission was to prevent anything like 1951 happening again.

A big avalanche is confirmation of a big snow event, but mainly it's an event in itself. You can look at it as a festival of physics bringing snow to life; or as one of the worst design faults in the natural world. Or as both at once.

The essence of the design fault is that snow cannot be relied on to stay where it falls. Combine enough of it – a cumulative depth of three metres, say – and a slope of more than 30 degrees, and the chances are that sooner or later it will slide. Rocks on the whole are different. If held in place by ice they will move when it melts. Otherwise, after an earthquake or volcanic eruption they find their natural angle of repose and stay in place. Snow moves because of gravity but also because of its changeability. What turns a placid-looking slope into an avalanche is usually a weak layer under the surface that makes the whole snowpack unstable, or an unusual layer to which new snow fails to bond. Either way, the secrets lie beneath. They have names like graupel, sugar snow and depth hoar, and the latter in particular can be a giveaway. Formed when rising water vapour turns compressed snowflakes into large ice crystals, it is bad at holding onto the snow above. A mountain guide who can sense depth hoar under his or her feet – or who takes the trouble to look for it in a systematic examination of the snowpack – can save lives.

In the 1860s, Joseph Bennen was considered one of the finest guides in Switzerland. He was, according to one avalanche historian, a strange and lonely man, but his clients revered him as almost superhuman. He lived with his mother and sisters in Laax in the upper Rhone Valley and was said to be at home in the mountains from there to Lake Geneva. In 1864 he led a party of six up the Haut de Cry, opposite Verbier, in winter.

Three of the six were local guides who seem to have goaded Bennen on against his better judgment. Close to the top they crossed a steep, snow-filled couloir and waited for him to follow. He did, but only after warning that he feared an avalanche. His client, a French climber named Phillip Gosset, brought up the rear and described what happened next:

> Bennen advanced: he had made but a few steps when he heard a deep cutting sound. The snow-field split in two some 14 to 15 feet above us. The cleft was at first quite narrow, not more than an inch broad. An awful silence ensued and then it was broken by Bennen's voice: '*Wir sind alle verloren*'. We are all lost.

Gosset plunged his alpenstock into the snow and leant on it with all his weight, hoping it might save him. "I turned towards Bennen to see whether he had done the same thing," he wrote afterwards. "To my astonishment I saw him turn round, face the valley and stretch out both arms." Gosset thrashed about and somehow kept his head above the surface but still had to be dug out of the bottom of the couloir with an ice axe. Bennen's body was recovered three days later from under eight feet of snow. He had died in a classic slab avalanche, released when the old snow could no longer hold the new. Bennen knew what the cutting sound and the crack meant. The slab would take them all with it. Rather than struggle he gave fate and the mountains a final, unforgettable salute.

The people of Galtur had no time for any of that. After the avalanche of 1999 it turned out that the fatal flaw in the starting zone above their village wasn't depth hoar, but a layer of snow that had thawed and refrozen several times in unseasonably warm weather at the end of January, before the February storms. The experts called this layer melt crust. You could equally think of it as a layer of rough ice. It wasn't weak. In fact, its strength allowed much more snow than usual to pile onto it before the whole edifice collapsed. And when that happened the melt crust helped it on its way. The avalanche set off like a 170,000-tonne bobsled.

Right at the top of that moving field of snow was the lightest, driest powder imaginable. These flakes had never been lower than 10,000 feet, never known temperatures above minus 10°C and had never been touched by anything but the wind and each other. There are avalanches that consist of this sort of snow and nothing else. One of them plays a stylish walk-on role in *Force Majeure*, the Swedish film about a young family on a ski holiday in France. The avalanche wafts over a mountain restaurant like a ghost, harming nothing except the relationship between Ebba, who cradles her children like a lioness, and her husband Tomas, who runs for his life (and then pretends he didn't).

Close to this in terms of solidity are the loose snow avalanches that trickle unseen off a thousand shaded faces after every snowstorm. At the other end of the solidity spectrum are strange rivers of old, wet snow that can advance down an otherwise bare mountain at the end of winter at little more than walking pace. These tend to be fed by snowfields higher up, disintegrating under strong spring sun. Anyone foolish enough to stand in the way will be crushed; anyone else can calmly take pictures. A crowd of Russian skiers can be seen doing just that in Youtube footage of several hundred tonnes of snow nosing slowly into a car park at the foot of Mount Elbrus in March 2018.

In the middle of the spectrum are the big avalanches that have everything: powder, loose snow and solid snow all together on a roughly synchronised rampage.

What the Swiss Institute for Snow and Avalanche Research (SLF in Swiss German) has shown is that the ferocity of the biggest avalanches may not be repeated for 300 years, and when it comes it is almost impossible to prepare for. The pressure in the middle of the Galtur avalanche will have been about 100 tonnes per square metre. (Imagine the weight of 50 cars on a manhole cover.) The powder blast that announced its arrival could strip the branches off a mature pine in a few seconds. The debris it carried hit people and property "like shots from a machine gun", and the snow it dumped could crush like a python.

I have to confess, I find this reassuring.

The reason relates to a bar of mixed repute high in the Tarentaise, in the holy of holies of French skiing. For most of the day this bar sits quiet and empty. For most of the night it heaves with the rich and beautiful and loud. I was never any of those things, but for one week in December 1982 I went there every evening with one of my brothers. He was 13. I was 16. We would go early, and we never spent a centime. We were there for the only thing that was free: the videos. We stood for as long as we dared in front of a large screen showing highlights of Warren Miller ski films and a selection of avalanches on a loop. It might have been the lighting but the loop seemed to have a blue tint, like Monet's *Winter on the Seine*. Its climax was a powder cloud billowing down from a range of semi-mythic British Columbian mountains and charging directly at the camera at

400 miles an hour. So said the narrator; he also said the camera was not remote-controlled.

The screen went abruptly black as the avalanche hit. We never knew what happened to the camera operator. We just solemnly assumed he'd made the ultimate sacrifice for his work. As for the 400mph top speed, I never questioned it. I simply stated to anyone who'd listen that that was how fast an avalanche could go.

The reason I find the SLF's work reassuring 35 years on is not that it proves the loop wrong. On the contrary, to within an order of magnitude it proves the loop right, especially if we substitute kilometres per hour for miles. It proves that avalanches can travel faster than a body in free fall.

The question is how. How does it move so fast, and how does the SLF know? The answers have been found in a valley not far from the Haut de Cry on the north side of the Rhône. The Vallée de la Sionne is deep, smooth-sided and austerely beautiful. No one lives here permanently because every winter the east-facing slopes avalanche with such force and regularity that they would flatten any normal dwelling. The slopes' average gradient is 35 degrees, which is perfect for avalanches. Shallower than 30 degrees and they would simply collect snow; steeper than 45 degrees and they tend to slough it off before it gets too deep.* The valley has therefore been given over to the SLF for research. Avalanches are its crop; its winter wheat.

---

* Alaskans will scoff at this. They say their maritime snow sticks perfectly well to slopes steeper than 45 degrees thanks to its high water content. There is no shortage of palm-dampening evidence to support this claim, although whole books have been written about the rescues mounted for those who put valour above discretion and ended up buried in snow that was not supposed to move.

On February 7th, 1991, a light snowfall in Whitehall hid markings left for an IRA van with a mortar bomb. The bombers missed their target in 10 Downing Street, *p.14*

Above: For Ernest Shackleton's Terra Nova expedition, snow came "like death in the night", p.19

Below: Sahara snow, Algeria, January 2018, p.27

Above: Wilson Bentley pioneered snowflake microphotography in Vermont in the 1880s, p.39

Below left: Bentley's snowflakes had to be photographed in haste, before they melted, p.39

Below right: Ukichiro Nakaya's snowflake classification forms the basis of snow science, p.42

low: "Tiny miracles of beauty" from Bentley's collection of more than 5,000, p.39-40

Opposite, top: Eric Shipton's ice axe next to the footprint he saw near Everest, p.57-8

Opposite, bottom: Steve Berry's "Migoi" prints on Gangkar Punsum in Bhutan, 2016, p.61

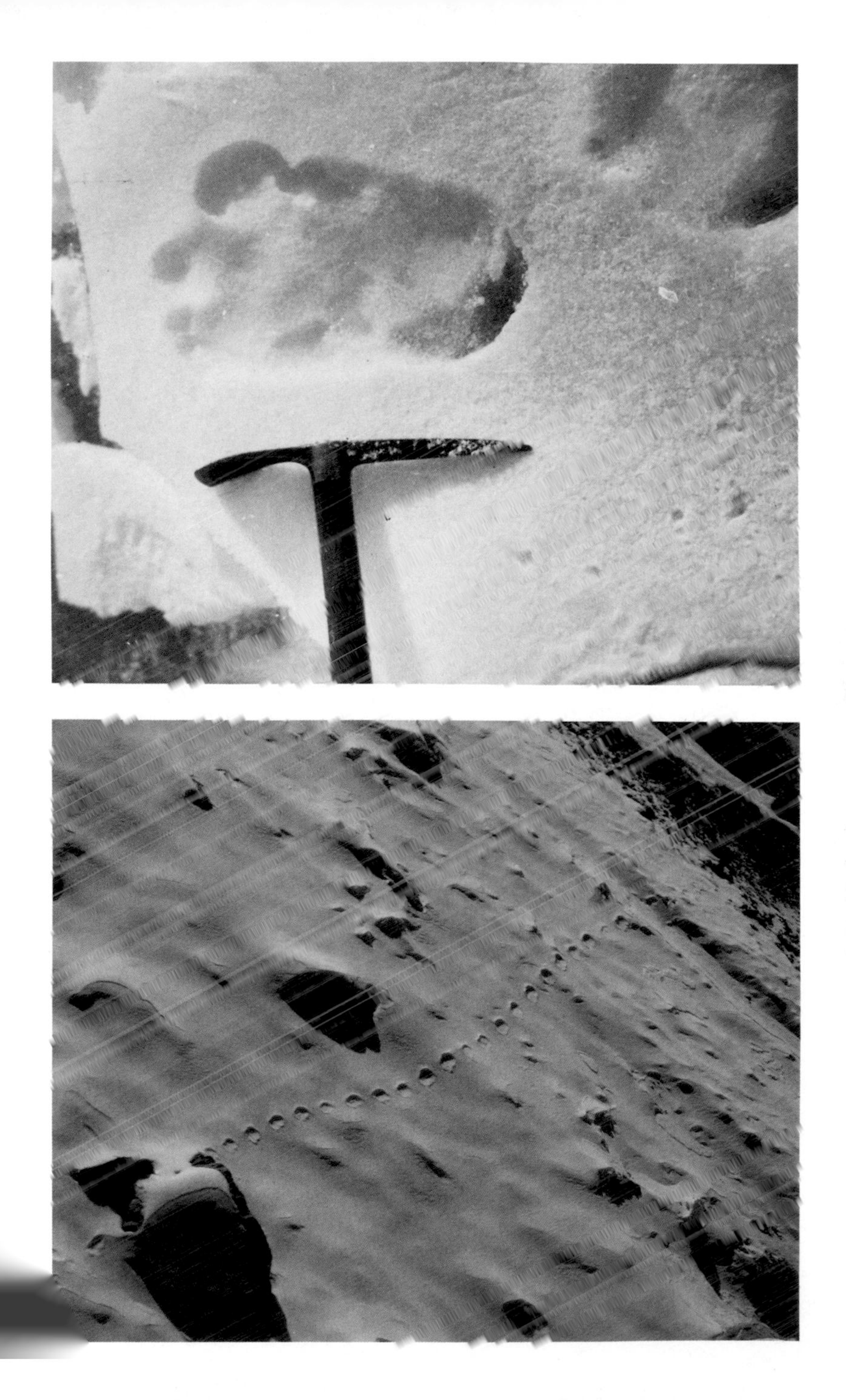

Above: Oddvar Brå's broken pole in 1982 has entered Norwegian folklore, p.67

Below: No longer in usable condition, the oldest known skis predate the pyramids by thousands of years, p.71

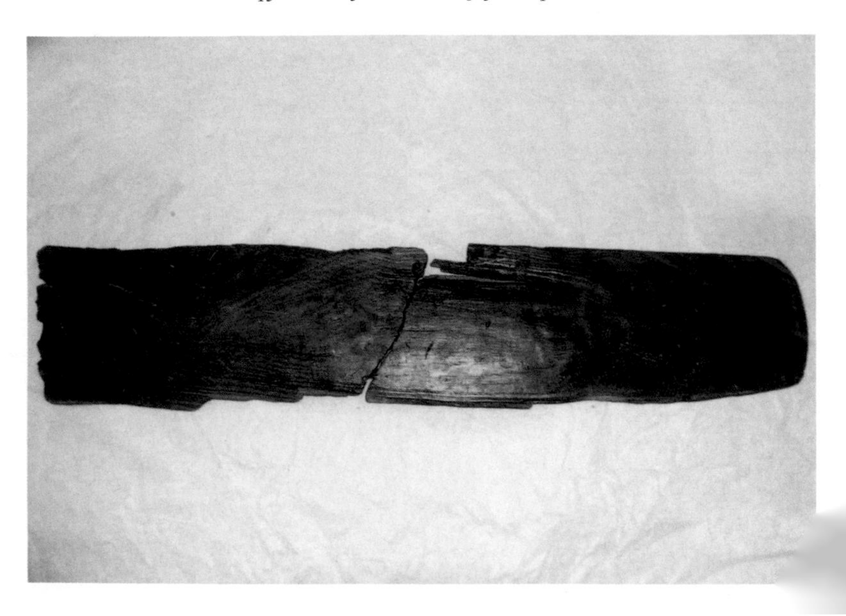

Above: 'Morning after a Snowfall at Koishikawa' by Katsushika Hokusai, p.94

Right: A winter scene from 'Les Très Riches Heures du Duc de Berry', manuscript illuminated by the Limbourg Brothers, c. 1416, p.87-8

Above: Detail from Bruegel the Younger's 'Census at Bethlehem', p.88

Below: Joseph Turner painted the 1808 avalanche at Grisons, but never saw it, p.86

Above: 'Snow at Argenteuil' by Claude Monet, 1875, p.92

Below: Lucas van Valckenborch (1535-1597) specialised in falling snow, p.89

Opposite: In the bitter winter of 1947 snow covered much of Britain for four months... p.106

Above: ...and broke records in Bolton, in Lancashire, p.106

Below: Snow is a rarity in 21st-century London, and never deep, p.100

Above: The Yuki-no-Otani snow canyon west of Tokyo attracts tourists every spring, p.130

Below: A powder avalanche pours off Ajax Peak near Telluride, Colorado, p.139-40

Above: Sverre Liliequist outruns a pair of avalanches near Zermatt, 2013, p.153

Below: Cameras rolling, Rick Sylvester launches from Mount Asgard in 1976, p.170

Above: Iain Cameron with Scotland's last two pieces of snow, summer 2017, p.180
Below: Forecast with extraordinary precision, "Snowmageddon" buried Washington DC under two feet of snow, p.203

A light grey scar is all that remains of the Bonatti Pillar, which collapsed in 2005 because of thawing ice, p.190

Above: A fohn wind can bring warm air and misery to one side of Mt Blanc... p.245

Below: ...and perfect snow to another, p.247

At 1,850 metres above sea level on the east side of the valley an array of flow sensors, temperature sensors, density sensors and aircraft-style pitot tubes are mounted on a hardened steel mast, replacing one destroyed by the big snows of 1999. Along with the new mast, the Austrian government paid for a thick concrete wall at the same location and fitted it with pressure pads to measure avalanche impacts broadside on. Fifty metres up from the bottom of the valley on the opposite side there is a concrete bunker bristling with Doppler radar guns to measure speed and density.

Two weeks before the Galtur avalanche the SLF used explosives to trigger the biggest avalanche it had ever analysed in the Vallée de la Sionne. When it hit the bottom of the valley it bounced up and covered the bunker to a depth of several metres, forcing the scientists inside to tunnel their way out. This was the avalanche that destroyed the mast, but enough data was recorded to serve as a model for Galtur.

The volume of snow dislodged in the starting zone was put at 50,000 cubic metres. That quadrupled as it accelerated down the mountainside, gouging into a deep base left by three large storms over the previous three weeks. The SLF's estimate that the Galtur avalanche doubled in volume on its way down was therefore conservative.

The Sionne powder cloud reached heights of up to 100 metres and speeds of up to 250 kilometres an hour, and the speed was a function of three things. The first was gravity, dragging the snow down faster as the slope grew steeper. The second was topography. As hundreds of thousands of tonnes of snow joined this flow on the upper part of the slope, the descending avalanche was funnelled into narrow chutes lower down. This is typical for the Alps, and the only way for all the snow to get through the chutes is to accelerate. The third factor was the SLF's speciality: avalanche flow dynamics. The purpose of all its gadgets is to peer inside the avalanche. What they saw was a base moving slowest, held back by friction. Above that and sliding forward over it was a murderous "saltation layer" of loose snow, chunks of snow, ice pellets and churning debris. And above that was the powder cloud, fed and pushed along by the saltation layer, eight times as dense as air and with nowhere to go but up and forward. All three factors were at work in Galtur too, acting on a much larger volume of snow that was sliding off a melt crust. That is why it went so fast.

Some experts distinguish between powder clouds and powder blasts, and with good reason. A cloud can be harmless. A blast – the airborne bow-wave of a big avalanche – cannot. Jill Fredston, one of Alaska's best avalanche brains, described the effects of powder blast at an avalanche to which she was called near Anchorage:

> The avalanche had clearly made a forced entry. Where there should have been a two-seater outhouse with brown walls and a green roof, white toilets sat on an exposed concrete pad. Nothing much bigger than a matchstick was left of the outhouse walls. In the surrounding forest, trees the diameter of basketballs had been yanked from the frozen earth like weeds… Some standing trees had pebbles embedded like shrapnel in their trunks, 40 feet above the ground. Picnic tables weighing 350 pounds had been tossed like Frisbees half the length of a football field.

Fredston says the dust cloud from the collapsing World Trade Center on September 11th, 2001, reminded her of powder blast from an avalanche. In terms of physics it is similar to the pyroclastic flows of hot gases down the flanks of erupting volcanoes, but cooler.

Not many people know what it is like to be inside a powder blast, but Colin Haley does. He lives in Seattle but is at home wherever there are mountains. In April 2015 he was acclimatising for a climbing expedition in northern Nepal when a 7.8 magnitude earthquake shook the whole country. Its impact was random, and all things considered Haley was lucky. When the quake struck he was with his climbing partner in the village of Kyanjin Gompa in the Langtang Valley, near Tibet. Within a few minutes Langtang itself, a collection of tea houses and stone dwellings five miles down the valley from Kyanjin Gompa, was buried under 40 million tonnes of rock and ice. A large part of the mountain above it had shaken loose and crashed to the valley floor.

Kyanjin Gompa was spared the rockslide. There was time for most of those living and staying there to run outside and make for a grassy plateau east of the village. That's what Haley did, thinking it the obvious safest place. He described the coming of the blast in his blog:

> I was just about free of the last buildings, almost into the grassy meadow, when I glanced back and saw the avalanche. No one had heard it beforehand, because there was so much noise of collapsing buildings and people screaming. No one saw it beforehand because of the thick cloud ceiling not far

> above the village. I saw a humongous cloud of snow descending through the cloud layer and down-valley towards us… I have seen plenty of big avalanches in Alaska and Pakistan, and none of them were anything like this. The avalanche that was coming down through the clouds and across the moraine seemed to be 300–400 metres tall. It was merely the powder cloud, the actual debris of ice and rock having stopped on the uphill side of the moraine, but [it] was moving much faster than I've usually seen in powder clouds.

Haley shouted to alert others to what he had seen, and started running:

> The avalanche was upon me about halfway across the meadow, and I crouched down on my hands and knees… The wind was incredibly thick with snow, and I pulled my hood down past my face, creating a little pocket in which to breathe. Within a few seconds there was so much snow in the air that it blocked out all light… Now I was scared… despite desperately trying to stay in place, I started to get pushed across the meadow, still crouched on my hands and knees. I had only been pushed for a couple of seconds and a few metres when the real blast hit. I guess it could be described as a pressure wave. I was hit by an incredibly powerful blast of wind, completely in another league from the worst winds I've ever seen in Patagonia, and in an instant I was airborne… I was just rag-dolling through space, and it felt very violent. I definitely thought, "OK, this is it. This is the avalanche I die in."

Haley estimates that he was thrown 30–40 metres, landing on his head near the bottom of a steep slope below the plateau. He suffered injuries to his neck, but nothing serious enough to stop him continuing to run. He was terrified of more avalanches. The powder cloud alone had picked up anything not made of stone and thrown it around like litter. In about a minute it had covered the entire valley in a foot of snow. Eventually Haley realised that if more avalanches did come he would not be able to outrun them. "And with some reluctance," he wrote, "[I] started walking back towards the village."

Humans caught in the main body of an avalanche tend to die in one of two ways. They get beaten to death, or they suffocate.

Once, on a press trip to Chamonix, I was told I should meet an extreme skier from Oregon with a reputation for great style and easy banter. His name was Dave Rosenbarger, "American Dave". He was downstairs in a bar on the Rue Whymper while we – an odd bunch of journalists and bloggers hoping for a glimpse of real-life superheroes in action – were upstairs having dinner. But I never did meet Dave because I turned in early that night and by lunchtime the next day he was dead. He had driven through the Mount Blanc tunnel with some friends and got caught by an avalanche in the throat of a couloir high on the Pointe Helbronner. He was not buried; he was broken, dying from his injuries in a helicopter on the way to hospital.

In death as in life, American Dave was exceptional. Most avalanche victims die gasping in an airtight coffin formed naturally around them. The terror is unimaginable, but the physics is fascinating.

The first thing snow does when it moves is sinter, as if sat on or rolled into a ball. Any flakes that still have their original shape lose it, and the fine six-pointed microstructures of stellar dendrites are the first to go, squashed in an instant. Dr Percy Bartelt of the SLF explains: "You might start with nice fluffy snow but when you shake it it breaks up into granules. These granules become hardened and compacted; they have approximately the density of wood. Then –" and this is the fascinating part "– the avalanche dissipates its potential energy into kinetic energy but also heat energy." A fleeting temperature increase deep in the avalanche leaves pockets of moisture on the skin of its billions of compacted granules, lubricating their movement in relation to each other. "Then, after the avalanche, the granules quickly refreeze, creating a very hardened material. This all happens in maybe two minutes."

These three processes – granularisation, fluidisation and refreezing – create a perfect, instant custom fit for anything trapped in an avalanche. They are the reason why even a victim who has the presence of mind to bring her hands to her face before coming to a stop in the hope of creating an air pocket may not be able to as much wiggle her fingers having done so. They are the reason Dr Bartelt says his favourite tool for digging through the snow pile left by an avalanche is a chainsaw.

Phillip Gosset, after the avalanche on the Haut de Cry that killed his guide, wrote that he tried to keep his head out of the snow with a treading water motion. When that failed he tried to protect his head with his arms, but the pressure "was so strong that I thought I should be crushed to death". Once he was buried he found he couldn't uncover his head because "the avalanche had frozen by pressure the moment it stopped and I was frozen in". His repeated mentions of the pressure are a reminder that moving snow can kill simply by squeezing, but "frozen by pressure" glosses over the real irony of death by avalanche: it's the law of conservation of energy – potential to kinetic to heat – that creates the vacuum that snuffs out life. Nature can be cruel.

After the Winter of Terror the Swiss and Austrian governments vowed never again to let their mountain towns be overrun by avalanches. Before 1951 the only serious work on avalanche behaviour had been done in Russia, where whole prison camps were buried in the Khibinyi Mountains near Murmansk and on Sakhalin island in the Far East. Dalstroy, the dreaded Gulag administration, wanted to know why. The answer was that it was easy to underestimate the threat posed by a given avalanche path without historical records or a good understanding of what made it an avalanche path in the first place. But Soviet research did not get much of a hearing in the west in the age of Stalin, so it fell to a materials engineer from Dübendorf, a Dr Adolf Voellmy, to do the work for Europe.

It was Voellmy who produced the first mathematical model for predicting how far an avalanche would reach on the basis of the volume of snow in the starting zone and the gradient below it. The Voellmy model appeared in 1955 in a paper titled *Über die Zerstörungskraft von Lawinen* (On the Destructive Power of

Avalanches). It was surprisingly little noticed at the time, possibly because it was a private commission for an Austrian construction firm, but it was later recognised as a landmark piece of research. It was used to transform building codes and justify the hundreds of miles of avalanche barriers that nowadays stripe the snowfields of the Alps like furrowed brows. But it failed to predict Galtur, which is why, since the turn of the century, the SLF has revised its calculations for each of the 800,000 avalanche paths on its avalanche hazard map of Switzerland. The idea is to let those living below them know if they would be at risk from a once-in-300-years event. The assumption is that one day, possibly in 300 years' time, possibly tomorrow, that avalanche will come.

Climate change will have a big say in the matter. Over the centuries avalanches have been good indicators of snowy years; you can't have one without the other. If warming weather systems mean that snowstorms could get more intense before they all turn to rain, as the Clausius-Clapeyron equation suggests, the same may be true of avalanches. They could get even more powerful before fading into history.

In the meantime, *Homo sapiens* tempts fate. Avalanches may kill 200 people in an average year but statistically, on a planet of seven billion, they are still something that happens to other people. That has been my experience, at any rate. In 1992 I strained my lungs to climb to nearly 6,000 metres on a beautiful Kyrgyz mountain called Khan Tengri. I stopped to admire the view at a place where everyone did, even though it was in the middle of a well-known avalanche path. That day nothing moved. Exactly a year later, from exactly the same place, four climbers were swept off the mountain. No one thought to change the route, and ten years later 14 more were killed there.

The victims are not widely remembered, but the survivors are. Skier Sverre Liliequist is remembered for his descent of a snow-laden face a short helicopter ride from Zermatt in the spring of 2013. The occasion was a team competition between Europe and America. Liliequist was an organiser of the event and he was keen for it to shine in front of the cameras. He is a former World Cup racer, confident in his technique and built like a running back. Half way down the face on his second run of the day he checked his speed for a fraction of a second above the first of two big jumps. Something, perhaps that movement of his skis, released two large avalanches, one either side of him. Each one started in a chute that channelled its flow while Liliequist barrelled on down, unaware of the danger, and launched himself into the second jump – a huge, exuberant backflip off a cliff the height of a house.

At this point you wonder: is it only as his back arches in the air and his eyes start looking for his landing zone that he sees the snow racing to catch him from both sides, in two furious white tiger claws, upside down?

That's what I wondered, anyway. When I asked him he said he didn't see the avalanches even then. "There was a lot of loose snow that day from previous runs. Of course this was a little bit extraordinary but I didn't find out until I turned round at the finish line and looked back up. On the jump I wasn't aware of the amount of snow. I was sticking to my plan."

It doesn't look planned. It looks like a might-as-well-go-out-in-style moment of pure suicidal chutzpah. It looks as if there is only ever a tiny chance that he will pull it off. Crazy odds, but he'll take them. Rotational momentum brings his skis around over his head. He nails the landing. For a moment it looks as if the avalanches are going to merge and bury him, but he skis out from under them like a magician.

CHAPTER EIGHT:
AT PLAY IN THE SNOWFIELDS OF THE GODS

Log cabin girl: But James, I need you.
Bond: So does England.
[Exit Bond in yellow one-piece, holding skis]
Log cabin girl: He has just left, he has just left.
The Spy Who Loved Me, screenplay, 1976

Without an avalanche a backflip is nothing nowadays. A hurricane, invented for World Cup aerials contests, is a quintuple twisting triple somersault. A backside triple rodeo, invented for the Winter X-Games, is a triple back somersault with the rotation at right angles to the direction of travel and a final twist on landing. The Winter X-Games is an annual carnival of daring in which young athletes update their social media profiles hanging upside down over enormous groomed snow jumps in Aspen. The Freeride World Tour is the same sort of thing, off piste.

All these are by-products of some of the most expensive distractions ever devised by humanity. All depend on the invisible fluid layer on snow and ice crystals that makes them slippery.

They are commonly known as winter sports, but that doesn't do justice to their extravagance or their exuberance. Gaze on them and the question arises like a cloud from an inversion layer: why? Why so many billions spent? Why so many millions hooked?

There are plenty of sober answers. Humans in winter get cabin fever and need to get out. Professional skiers lose their edge and need to find new ways to compete. Television companies need ratings, and the thrills made possible by that fluid layer deliver them.

Alternatively, you can trace it all back to Bond. To mention his name in 2018 is to invite eye-rolling and ridicule, much of it justified. But not to mention Bond in a book on snow would be a crime. No one in the modern age has used snow for escape and escapism as effectively or as contagiously as 007. Ian Fleming, and especially those who bought his film rights, knew nothing would spirit their audience away from the smog and strictures of post-war suburbia as quickly as skis* on snow. Even in the 60s the combination was still as exotic for most people as jet planes. It was also perfect for getting Bond quickly in and out of scrapes, and thereby hangs a tale that starts, oddly enough, in California.

Until 1955 Squaw Valley was an unspoilt secret of the Californian Sierras. A mile and a quarter above sea level and a short

* and sex, of course.

drive from Lake Tahoe, it was home to thick stands of black oak and Douglas fir. On the floor of the valley lay a beautiful grassy meadow that often flooded in the spring. Every summer it was filled with wildflowers.

A rough, unmetalled road headed up one side of the meadow and petered out where the mountains took over. In 1955 there was a rope tow, a chairlift and a single ski lodge recently rebuilt after burning to the ground. And that was it. Then, in June, a group of American businessmen travelled to Paris to tell the International Olympic Committee that 35 feet of snow fell in Squaw Valley every winter.

The group was led by a tall, ginger-haired East Coaster by the name of Alexander Cushing. He'd been to Harvard law school but that did not mean he held the law in high esteem. What he said about Squaw's snow wasn't true, but it didn't matter. No one on the IOC had been there, which was what the Cushing delegation hoped to change. Their goal was to turn the place into a huge resort and spread the idea of California as a place of snow as well as sun. Wooing the Olympic Committee was a means to that end. "I had no more interest in getting the games than the man in the moon," Cushing said later. "It was just a way of getting newspaper space."

He got plenty of that. He had brought with him an outsize papier-mâché model of the resort he planned to build. The model was too big to be installed at the IOC's Paris offices, and so it was given its own premises.

"People really wanted to see this thing," an old acquaintance of Cushing's recalls. "It was a few blocks away and in those few blocks Cushing had a chance to schmooze the committee members. He was just like the devil, the most charming of gentlemen. He'd lie to your face. Not as bad as Donald Trump, but you have to give him credit."

On the question of snow the committee members took Cushing at his word. Squaw Valley beat Innsbruck by 32 to 30 in the final vote and within five years $80 million had been spent on roads, bridges, ski jumps, ski lifts, ice rinks, hotels and the first ever Olympic village.

Teams from 33 nations booked passage to the first Winter Olympics in the American west, and Walt Disney was hired to stage the opening ceremony.

There was only one problem. No snow. Not even a hint of it. When Disney flew up to Squaw from Los Angeles early in January 1960 the roofs of all of the brand-new buildings were still brown. The roads were black. The trees and slopes were bare.

In any normal year Squaw should have been at least six feet under by January. It sits in a horseshoe of peaks that reach up and force the moisture out of every storm approaching from the west. Several fronts should have come through by the time of Disney's visit, but none had. He and Cushing faced the prospect of humiliation, but neither man surrendered easily to fate. It was Disney, after all, who said if you can dream it, you can do it.

The two men asked around for miracle-workers. Someone suggested the snow dancers of the Western Shoshone nation, whose tribal lands once covered half of Nevada. The snow dancers who could call on folk wisdom accumulated in these parts over more than10,000 years. In particular they knew not to take the sky for granted. They had ways of showing nature a little respect.

Disney decided he had nothing to lose. The call went out and the dancers arrived from Reno and Carson City and dozens of smaller settlements scattered across the Great Basin between the Sierras and the Rockies. They put on costumes of deerskin and rabbit fur and stamped and chanted in a wide circle in Squaw's brand-new town centre. And still it wouldn't snow.

Disney asked around again. This time the name of Irving Krick came up. Krick was a meteorologist and, like Disney, a showman. In 1944 he had served on a team of American forecasters assigned to help General Dwight Eisenhower with the timing of the Normandy landings. His method was bogus, predicting the future based on the past rather than on current observations, but he claimed credit for D-Day anyway.

Krick had a neat moustache and a fondness for expensive knitwear. After the war he decided there wasn't enough of a future in conventional meteorology for someone of his style and energy, so he switched from forecasting the weather to making it. Scientists were experimenting for the first time with the seeding of clouds with silver iodide to precipitate precipitation, and he turned the idea into a business. Irving P. Krick & Associates could make it rain or snow. All you had to do was ask and pay. By 1960 he was turning over more than a million dollars a year.

Disney hired Krick as the Olympics' weather engineer, and Krick's people installed a ring of silver iodide generators around Squaw Valley. At first all they could do was wait. The skies were clear; there was nothing to seed. But on January 10th clouds rolled in from the Pacific and the paraffin flames under the generators were lit. For a day or two nothing happened, but the weather kept coming and the temperature fell, and flakes the size of cucumber coins started drifting earthwards as they should have been for weeks.

Three feet fell in the valley; seven on the slopes above. And then it rained. A wind shift brought sub-tropical air up from the south and most of the snow disappeared. Eventually, with hardly any time left before the world arrived to see Squaw's famous snow, the cold returned. Disney's opening ceremony unfolded in a blizzard.

Naturally Krick and company said it was all their own work. So did the Shoshone. There was a third possibility – that Disney got lucky – but that is less alluring than the idea that if you want snow badly enough, you can make it come.

Two years after the Squaw games, snow dancers from the Southern Ute tribe were invited to perform for the opening of the new ski resort of Vail 600 miles to the east in Colorado. As they danced, it started snowing. They returned in 1999 for the World Ski Championships, and again in 2012 to see if they could help end a drought. Both times, they brought snow.

The Native Americans of the intra-mountain west may know something the rest of us don't; something passed down the ages with the legends of the storm bird and the wendigo. Or they may just know from long experience to accept snow dance invitations only when the sky looks promising.

"We should do pretty well here from now on," Cushing told *Time* magazine shortly before the games, and he for one did very well indeed. The games were a resounding success. They put Squaw on the map as a cross between California's Chamonix and Hollywood-in-the-sky (also known, inevitably, as Squallywood) and they made Cushing a substantial fortune.

He was a maverick who attracted others, and one of those who drifted up to Squaw in the late 60s was a skinny, thoughtful 25-year-old called Rick Sylvester.

Sylvester never thought of himself as a stunt man. It sounds too professional, and he is an amateur *par excellence*. In one sense this is irrelevant – he did what he did – but in another it may be illuminating. Perhaps a professional would never have dared. Perhaps it took an amateur to pull off the most outrageous stunt in the history of cinema.

Born in Brooklyn, Sylvester attended Beverly Hills High School as a teenager and then the University of California at Berkeley. He majored in sociology and Scandinavian literature and language at the high noon of the counterculture. He was radicalised by the 60s, but instead of revolution he took up climbing: he was one of the early wall rats on El Capitan, the 3000-foot granite citadel in Yosemite National Park. He also learned to ski in the resorts around Lake Tahoe. One instructor he admired, Jim McConkey, was well known locally and his name is worth remembering.

Sylvester arrived in Squaw in 1967. One reason he stayed was that he liked the snow. "I've skied champagne powder and I actually haven't enjoyed it," he says down the phone 50 years later. "Our snow has more body. It has more resistance. We like it." He likes the rituals of living with it, too, and he measures his senescence by his ability to shovel it.

He stayed in Squaw, although not in the sense of never

leaving; wandering souls need to wander, after all. In 1969 he taught climbing in Switzerland for a summer and in the autumn of that year he went to Hawaii to earn some money in construction. It was there, he says, that he started having dreams – real dreams, not daydreams – about jumping off El Capitan on skis. Climbing all the way up it was still a rare accomplishment. Falling down it on purpose, even more so. It had only been done once, by a pair of Californians who jumped in summer (they deployed their parachutes successfully but still broke bones on landing).

Sylvester's dreams were set in winter. In the end he skied off El Capitan not once but three times. He claims, half-joking, to have self-diagnosed obsessive-compulsive disorder and to have become fixated on getting the perfect camera angle for the jump. He never did. His main technical challenge was to remove his skis so that they wouldn't pull him into an awkward position and get tangled in his parachute. For this he had devised a system that worked well in practice. It was based on an unusual binding, the Spademan, designed by a San Francisco orthopaedic surgeon of the same name who travelled up to Squaw to demonstrate it for the local ski patrol (of which Sylvester was a member). Unlike most bindings it had no toe-piece. Crucially, it was released by pulling up on the heel clip, not pushing down.

Sylvester's system was rudimentary. "I attached some half-inch climbing webbing [to the heel clip] and ran it up to my knee," he said. "I had a piece of Velcro sewn on the outside of my ski pants just above the knee on both sides, and at the top of the webbing I had a piece of wooden dowel with a piece of

Velcro attached to that as well. So I only had to reach to my knee on both sides to release the bindings."

The idea was to save half a second by not having to reach right down to his heels, "which, who knows, might make a difference when you're literally flying to your death unless you do something to arrest that".

The first jump was filmed by a gonzo crew making what they called ski porn. Sylvester was half way down the face before he had removed the skis and by that time he had been out of camera shot for several seconds. "I might as well have skied off a 20-foot drop," he wrote later in the *Tahoe Winter Times**. The second jump was perfect but Sylvester decided afterwards it should have been filmed from a helicopter. The third *was* filmed from a helicopter, but the camera malfunctioned and he nearly died falling from a tree on landing.

There his attempt to film what he had conceived as the greatest ski jump on Earth might have ended. He had tempted fate and survived. He had stirred up a little notoriety on the front pages of the *San Francisco Chronicle* and the *Los Angeles Times*. More importantly, he'd made a name for himself where it mattered, in Squaw Valley. Cushing knew who he was. Fellow skiers

---

* There is another version of this story, told by the director of the film crew trying to shoot the jump, Mike Marvin. In this version Marvin invents Sylvester's ski release mechanism, not Sylvester, and carefully schedules the jump for 11am only for Sylvester to lose his nerve for 24 minutes at the top of El Cap. By this time the helicopter, worried about arousing the suspicion of the National Park Service, is out of range and misses the shot. The only thing both sides agree on is that they had a falling-out. The version told here is Sylvester's, on the basis that even if he did pause to reflect for a few minutes at the top of the cliff, which he insists he did not, it would have been an entirely reasonable thing to do.

and climbers might even point him out across the parking lot. He'd never competed in the Olympics but he had earned some stripes.

And then the phone rang.

It was Albert "Cubby" Broccoli, producer of the Bond films, on the line from London. A photograph had been brought to Broccoli's attention of Sylvester airborne on skis above a snow-scape of towering cliffs and otherworldly glaciers. The picture was a Canadian Club whisky ad. It was a fake – a genuine frame from one of the El Capitan jumps superimposed on Baffin Island in the Canadian Arctic. Worse still, it caught Sylvester in a terrible silhouette. Bum out, legs rigid, hands clawing at the air. It radiated fear. But Broccoli had also seen some footage of the jumps and something about them seemed worth replicating. By Bond standards Broccoli's last effort, *The Man with the Golden Gun*, had been a dud. (Not incidentally, it lacked snow.) The next film had to be big. Would Sylvester do the jump again?

He had just signed on for unemployment benefit at the end of a season as a ski instructor. The money being offered for the jump was good. He has never said how much he was paid, but it was five figures then and would be six now. He said he would.

When Sylvester was asked where he thought the stunt should be shot, he suggested Mount Asgard. From Scandinavian literature he knew the Asgard of legend was the home of the gods. From climbing he knew it as a semi-mythical place to go when someone else was paying.

Mount Asgard is so far north that the snow on its summit is completely reliable even in summer. Elsewhere on this very unusual peak, snow is virtually non-existent even in winter because it has nowhere to land. The mountain has two summits, and each one is guarded by vertical walls of granite nearly a mile high. They sit 25 miles north of the Arctic Circle, rising from a moat of glaciers like the twin pillars of Valhalla.

The sheer drop from the top is 4,000 feet, which is a thousand more than from El Capitan or an extra five seconds at terminal velocity. The background, all snow and rock, is more like Austria than Yosemite. This suited Mr Broccoli, given the locations already suggested by his screenplay. In the foreground there is a short, steep slope to the abyss.

The nearest settlement is 50 miles away in Pangnirtung, at the entrance to a mighty fjord. In July 1976 Eon Productions, Broccoli's company, arrived there with a full second-unit crew, a Bell Jet Ranger helicopter and a bright-yellow one-piece suit for Bond, chosen by Sylvester from the Willy Bogner range. It was offset by a scarlet backpack for his parachute. The crew was billeted in the only hotel in town, the Auyuittuq Lodge, and they feasted each night on Arctic char.

For a week the weather held. Each day the crew flew up the Turner Glacier to Mount Asgard to scout camera positions and hurl smoke bombs off the mountain. The idea was to see which way the wind would blow Sylvester, but he was nervous about wasting such a long sequence of clear days so far north. Sure enough, the moment they were ready the weather broke. "It started as Scottish mist and then got worse," he says. "Things were looking bad."

Eight hundred miles to the south, the Montreal Olympics were under way. Sylvester calls himself an Olympics addict and the lodge had recently acquired satellite TV. So he was able to get a daily fix, but it didn't distract him much from the fear seeping into his soul:

> I was getting superstitious. I come from a period of amateurism but this was filthy lucre. It was for the money. It was like a curse. As each day remained bad I found myself secretly rooting against the project, having melodramatic thoughts. 'I don't have to do it today – another day of living.' I never said this to anyone, of course. It was too damning, but I was rooting against the thing happening despite all the time and effort and expense.

Apart from anything else, he was troubled by not being a pro. He was among professionals but wasn't one of them, either in the film business or as a skier or a stuntman. He was the eccentric willing to risk his life for money and a few seconds of film. He did know more about mountains than the rest of the crew, though. He told them about microclimates – about how just because the weather was miserable at Pangnirtung it didn't meant it was necessarily miserable at Mount Asgard. And so a new routine was started: the helicopter would fly two trips a day to the mountain to check on conditions there, once in the morning and once in the afternoon. This went on for nearly a week of Arctic char and darkening moods, lightened only by the Olympics. One night Sylvester overdosed on them. He slept in and woke to the worst weather yet. The morning flight was a washout. "But then a few hours later the afternoon flight goes out and the guy comes back and says 'come on, we can do it. It's clear'."

At this point Sylvester is fearful and confused. On the face of it his theory about microclimates has been proved right, but he has doubts. At the start of the trip the unit production manager took him on one side and promised earnestly that his life mattered more than anything. If he didn't want to do the jump, for any reason, all he had to do was say. But since then London has been calling. Has he done it yet? When's he going to do it? What's the hold-up? Sylvester wonders if the unit production manager is feeling the pressure and letting it affect his judgment.

The flight takes less than an hour and it's quickly clear the job is on. The colossal granite towers of Mount Odin and Thor Peak, mere tasters for Mount Asgard, are out of the cloud. Mist still swirls around the base of Asgard itself but the whole north face is ready for its close-up. Sylvester can take as long as he likes; at this time of year the sun will hardly set.

Two cameras are fixed to the mountain – one near the edge of the cliff with a side view of the jump, the other on a ledge directly under the point where Sylvester plans to launch himself. Their operators take up their positions, securely fixed to the mountain in the case of the man on the ledge. John Glen, the second unit director, is next to the camera with the side view. The master shot will be from the helicopter.

Sylvester side-slips to the edge on a belay. The snow is breakable crust and he wants to smooth it out a little so he doesn't catch an edge. "I may be dumb but I'm not stupid," he says. "Or is it the other way around?"

The Jet Ranger takes off and hovers over the glacier north of the drop. Glen asks everyone in turn if they're ready.

"I'm the final one he asks," Sylvester remembers. "He says, 'are you ready?'"

There are two things Sylvester has to get right: the ski release and his body position. He has to be flat and facing down before he pulls the chute. He's practised the release many times, but not in this suit or with heavy Lange ski boots, or off this cliff.

"Are you ready?"

Sylvester doesn't want to say yes but can't come up with a reason not to. "So I say yes, and he tells the cameras to roll and says to me 'go on, go ahead'." And he's off, the only one not roped to the mountain as if a typhoon is about to hit. It's a straight shot, no turns, no hesitation. After a few seconds he dumps his poles and then he's in space, gravity not yet going to work on him, heading straight out over 800 metres of nothing.

Forty years on the wide shot of those few seconds, with a soundtrack of high, keening violins (from a score by Marvin Hamlisch) and the realisation of the appalling scale of the drop and the fact that this has not been faked, still puts you in a prime seat at the Coliseum every time, staring jaw agape at a gladiator facing death.

Sylvester would write that once rid of his skis he "experienced some trouble getting stable". In fact he lets the wind blow him into a lazy back somersault. Then his boots drag him into a standing position for an agonisingly long moment before he can get face down to release his parachute. The view of the glacier below is partly obscured by mist, making it look as if he's arriving from the stratosphere.

How do you top that? Make the parachute a Union Jack. *The Spy Who Loved Me* was released in the year of the Queen's silver jubilee, and for many it was the flag on the chute that brought them to their feet. For me it was the jump, and the long, long silence of the fall. Compared with all the more complicated snow stunts since then it had a simple purity, never to be beaten. It was definitive. It was death-defying and life-affirming. The fact that it turned out to have been performed by a self-deprecating amateur with a degree in sociology and Scandinavian literature only made it more so.

The Mount Asgard jump was the apotheosis of Bond on snow. *Newsweek* called it the greatest stunt in the history of film. It nearly didn't make the cut, though, because the helicopter missed the shot. It captured the take-off run but little else because Sylvester fell so far, so fast getting into his stomach-to-Earth position. He was chiding himself about this almost from the moment of landing, and has been ever since.

After the jump the crew returned to Pangnirtung, where the aerial camera operator broke the news. The side-view cameraman was more optimistic. He thought he'd captured the whole thing. No one could be sure because they were shooting film that had to go to a lab before being viewed, and the nearest lab was in Montreal. The film was flown there and the news that came back was good. The second camera had enough. It even showed one of the skis hitting the parachute as it opened. There was no need for a retake, which was an immediate relief for

Sylvester, and an even bigger one in retrospect because when he repacked his parachute he repacked it wrong. There would have been a malfunction, he says. He knows this because there was one when he used it next, in California. He would have had to cut away the main chute and use the reserve, as he did in California, and the reserve was not a Union Jack.

Glen and most of the crew returned to England. Sylvester and a climbing buddy stayed in the Arctic for a few days. Having pulled off the trick of getting someone else to pay for them to go to Mount Asgard, they tried to climb it. The weather wasn't good, though, so they hiked half way back to Pangnirtung and caught a boat the rest of the way down the fjord.

The following year Sylvester attended an early screening of the film with his mother in Los Angeles. The story he tells is that at the end of the opening sequence a woman whispered to her friend in the row behind them, "How do they do that?"

Friend: They use dummies.

Mrs Sylvester, turning: Yes, my son, the dummy.

Back in Squaw Valley, Sylvester got a little more respect. Among his admirers was the son of his old ski instructor, Jim McConkey. The son's name was Shane. He was eight when the film came out, which is a good age to see it for the first time, and he went on to become an international extreme-skiing superstar, commanding and revelling in the spotlight as Sylvester never did. His signature jump was a double forward somersault off a cliff, keeping his skis on to enable him to open his parachute earlier and leap off lower cliffs than El Capitan or Mount Asgard.

As Sylvester remembers it, the first time the two of them

actually met was in the Squaw Valley post office, where they were picking up their mail. McConkey asked about Sylvester's quick release method with the webbing and the dowel and the Velcro. Sylvester had come to call it the fuddy-duddy method, and McConkey did sometimes use it. One of those times was in a spoof of the entire ski chase from *The Spy Who Loved Me*, complete with preposterous studio close-ups of McConkey as Roger Moore. Another was a jump off the Piz Pordoi in the Dolomites for a commercial sponsor. Piz Pordoi looks as formidably high as Mount Asgard but in fact is on a different, smaller scale. The drop is vertical and spectacular but only 1,500 feet. McConkey had trouble releasing his skis. He managed in the end, but by then had been free-falling for 12 seconds. By the time he flipped over into the stomach-to-Earth position it was too late. He hit snow before the parachute opened and died instantly.

In extreme skiing as in other forms of entertainment, the show has to go on. Partly in tribute to him, McConkey's friends made sure it did. In 2015 his closest friend and collaborator, JT Holmes, jumped off the north face of the Eiger for the CBS news show, *60 Minutes*. It was a jump with a twist: he ski-flew the first quarter of the mountain with the help of a parafoil that he dumped 300 metres before taking off. He then did two backflips before opening his chute. The descent was flawless – so flawless that he went back up and did it again, and the second time one ski wouldn't detach at the first attempt. There were no backflips, just a terrible wait for him to try again.

One of the CBS camera operators described it for the broadcast: "I see him flying in the air and I see one ski go but I don't see the other one. He had about two seconds when he couldn't get it off. I'm trying to film this but inside your heart's going, 'Oh Jesus, Oh Jesus, come on.'"

The ski came off and Holmes had time to open the chute, but his body language speaks volumes about his state of mind having done so. He hangs limp in his harness as if the adrenalin from cheating death is draining out of him. "It was every bit as intimidating as I thought it would be," he said. "It was a wild ride."

Holmes lives – where else? – in Squaw Valley. Sylvester runs into him occasionally and hopes sometime to be able to ask him whether, on that second jump off the Eiger, he had a moment to think "uh-oh, this is what happened to Shane". The opportunity to ask that question hasn't arisen yet. In the meantime Sylvester runs marathons in summer, skis his Sierra snow in winter and takes inordinate pride in shovelling it off his roof himself. "I figured success meant not just doing this thing but surviving," he says. He's 76.

CHAPTER NINE: LAST RITES

"But if they had no sun, they had snow. Such masses of snow as Hans Castorp had never till now in all his life beheld... The snowfall was monstrous and immeasurable, it made one realise the extravagant, outlandish nature of the place. It snowed day in, day out, and all through the night."

Thomas Mann, *The Magic Mountain*

Davos was once a by-word for snow. In a high, broad valley 100 miles from Zurich, it was the setting and inspiration for the wintry dreamworld created by Thomas Mann for Hans Castorp in *The Magic Mountain*. Later it came to host the World Economic Forum, which invites industrialists and idealists to commune in a conference shelter entirely insulated from nature. It is also home to Switzerland's Institute for Snow and Avalanche Research, and in 2017 the institute released a picture to accompany a new study of the prospects for snow in the country's ski resorts.

The prospects were miserable, and so was the picture. In the centre was the top station of Davos's Parsenn funicular railway. Next to it was a short stretch of ski slope. The slope was white but the mountain on either side was brown. Skiers were outnumbered by snow cannons – nine of them. The text by Dr Christoph Marty set out in clinical detail three possible scenarios for Swiss snow depth over the rest of the 21st century. The first was based on hope: a successful international carbon emissions limitation plan that halves emissions relative to 1990 levels by the end of the century. The next two were based on experience. One assumed rapid economic growth, stabilising global population and a mixture of renewable and fossil fuel energy production; the other assumed lower economic growth and a continuously expanding population. The study accepted the evidence of a link between emissions, temperature and snowfall, and then offered average snow depth projections for a range of elevations above sea level for the immediate future, the middle of the century and the end of the century. Its conclusion was blunt: "The projections reveal a decrease in snow depth for all elevations, time periods and emissions scenarios."

Under all scenarios the hardest-hit communities will be those between 1,500 and 2,500 metres above sea level; in other words, most of Switzerland's ski resorts. Even at 3,000 metres, and even with steep cuts in emissions, average snow depth will be halved by the end of the century.

The paper was not disinterested – Switzerland depends on snow – but its tone was dry and detached. It hinted only twice at human sentiment: once in its projections of future numbers of "snow days", which it defined as days with at least five centimetres of lying snow "because with regard to winter tourism this is the minimum snow depth to generate a winter feeling, build a snowman or go sledding"; and once in its title: *How Much Can We Save?*

There is no point mincing words. The way things are going the answer may be not much. Switzerland is saving snow wherever and however it can, wrapping thousands of tonnes of it in reflective white polythene to give it a better chance of surviving the summer and helping with the next season's cover. It is even wrapping glaciers. It is treating the symptoms of the retreat of snow because treating the causes would take teamwork with other countries that have less at stake. Until that teamwork happens, what used to be taken for granted will become scarcer and more precious. As it does, the debate about who and what is to blame will become even more rancorous than it is now. Businesses dependent on snow will become more and more resourceful or go to the wall. Businesses that make it will thrive. Citizens may simply adjust. They won't necessarily lose interest in snow. Some will, but others will become more interested in it. There may not be enough of it to slide down or land in after jumping off a cliff, but there will be enough to seek out for the pleasure of finding it. Snow will become treasure. We know this because it is already happening.

In the late autumn of 1933 a representative of the Scottish Mountaineering Club wrote to *The Times* with troubling news. A patch of snow at Garbh Choire Mór in the eastern Cairngorms had melted for the first time in 300 years. Written records chronicling the annual survival of the patch went back to the 1840s. For the years before that the club felt it could rely on the word of gamekeepers and stalkers on the great estates of

Banffshire and Inverness-shire, passed down through the generations from the 1700s.

That snow patch, at the foot of a cliff in a north-facing corrie that almost never sees the sun, was known to be the longest lived in Scotland. Some have tried to make the case that it was a glacier in the Little Ice Age but there is little evidence under or around it of the grinding that glaciers do or the deposits that they leave behind. For most of the past 10,000 summers it has almost certainly only ever been a patch. This has not stopped it becoming an object of obsessive interest for snow patch enthusiasts united by social media and a feeling for endangered species. They call this patch the Sphinx, after a rock climb above it, but they follow the lives of other patches too. Some years hundreds survive the summer; some years, nowadays, none do.

They are humble things, these patches, especially from a distance, which seems to be one of the reasons people like them. Another is that they can surprise. From across a valley they can look no more interesting than "a blob of ice cream", says Iain Cameron, an aerospace engineer who runs the Snow Patches in Scotland Facebook page as a hobby. Up close, they can retain a startling depth long into the summer. They become scalloped as they melt, like beaten armour, and in July they can reveal their bellies to the curious by way of tunnels big enough to crawl into. These start forming as meltwater trickles down the mountain underneath them. As warm air finds a way in they can grow into vaulted chambers filled with blue refracted light.

Since 1933 Scotland has been snow-free only in 1953, 1959, 1996, 2003, 2006 and 2017, and then only for short periods at the end of autumn. The winter of 2014–15 was one of the snowiest in half a century. That year 678 patches lasted the summer. But the frequency of full-melt years is picking up. As it does, so does public interest.

Cameron is used to being interviewed, and used to being asked what it is about snow patches that sends him off into the hills in search of them most weekends from May to November. He says it's a fascination "with how, somehow, against the odds, they can endure". But he's also interested in the science of melting. A patch ten metres deep, which might have been created by a falling cornice, will last not twice as long as one five metres deep but exponentially longer. Its ratio of volume-to-surface area is much higher and it might well survive even an unusually warm summer.

In 2017 the winter was not especially snowy and the summer was warm. By late September the Sphinx had shrunk to a pair of large plates, neither one big or strong enough to lie on. Cameron hiked up to pay his respects and was photographed holding one of them like a large silver salver. He said it would be gone by the next day, and it was.

I wondered if an interest in Scottish snow patches could travel. As a devout snow internationalist I hoped so, and it turned out that it could. There are a couple of other out-of-season snow phenomena that Cameron would like to visit, and one is the 1,700 year-old Kuranosuke patch on Mount Tateyama in Japan. This and two other Japanese patches have recently been given glacier status on the basis that deep within them they contain fossil ice as opposed to firn, but there are reasons to wonder if they truly deserve this promotion. One is that the fossil ice classification has been made by a "specially appointed" professor of stratigraphy at Shinsu University. The other is that this professor admits the patches are "somehow getting smaller".

The conventional wisdom on Japanese glaciers, for many decades, has been that they do not exist; that just because Japan is a very snowy country it does not mean it has to have them. The fact is that its mountains are low by Alpine standards. Its latitude is not especially high, and its summers are warm and windy – an extremely effective combination for melting.

"There are currently no glaciers in Japan," Professor Teiji Watanabe of the University of Colorado at Boulder has written. There are perennial snow patches, but only where drifting and avalanches add depth to in situ snowfall, usually in the lee of high crests in upland Hokkaido and the northern Japanese Alps. A total depth of 15–20 metres is needed to last the summer, and even Japan's remarkable snowfall cannot manage that alone.

Glaciers need year-round sub-zero temperatures to survive. Temperatures can and usually do rise above freezing in summer at their snouts, but in their accumulation zones it has to stay cold or there will be no accumulation. Average air temperatures in summer fall below freezing only above 4,000 metres in Hokkaido and 3,000 metres in the northern Alps, and try as they might, Japan's mountains do not reach that high. If their snow patches are getting smaller that would unfortunately fit a pattern. A two-degree fall in average air temperature would quickly turn them (and the Sphinx) into healthy, growing glaciers, but the reality is a rise of about half a degree since the war, and that is sweaty for a snow patch.

Kuranosuke will cling on for a while yet. If only the same could be said for the snowfields of the Rwenzori Mountains, which rise from a mantle of rainforest in the heart of Africa to the upper reaches of the troposphere. Ptolemy called these strange, cloud-wreathed peaks the Mountains of the Moon, two centuries before Christ and without ever seeing them. He was recycling an Egyptian myth that they sheltered eternal springs that were the source of the Nile, and for 2,000 years he was believed. In fact there are no eternal springs – but there are eternal snows, dizzyingly high. To reach them means leaving the rainforest behind and walking for a week through bamboo groves and an African alpine wonderland of giant heather and lobelia. There, if the clouds part, the snows allow those down below a glimpse.

The Rwenzori straddle Uganda's border with the Democratic Republic of Congo, 30 miles north of the equator, and they are the other place where Iain Cameron would like to see snow surviving against the odds. He will have to hurry. Beneath their snows lie real glaciers but they have lost 85 per cent of their surface area since first trodden by a European in 1906.

All six main peaks in the range used to be garlanded with glaciers. Now only the two highest still support them, and these have shrunk to coy flashes of white, like dog collars on priests anxious to blend in.

Snow will still fall above 4,000 metres in the rainy season, rising to 5,000 metres and higher if air temperatures go on rising, but the glaciers that this snow used to replenish will be gone by 2030 at the latest. This forecast is based on the rate of glacial retreat over the past century. It may prove alarmist, which would be wonderful, but the fate of high snows near the equator elsewhere in the world suggests the opposite.

In 1987, with a backpack and a friend from university, I headed north out of La Paz in the back of a Jeep, up an old dirt track much travelled by gringos in Bolivia. It led to Chacaltaya, the world's highest ski slope. The name is said to refer to the road itself, Chacaltaya meaning "cold road" in Aymara. But at the end of it was a glacier, a drag lift and a hut that all went by the same name. The hut stood and still stands at a lung-busting 17,400 feet. The glacier was no beast; it had never carved a gulch for itself, resting instead like a pancake on the slope, but it was a substantial pancake, perhaps half a mile long and quarter of a mile wide. The lift wasn't working, but that was because it was mid week in the dry season with no fresh snow, and we were the only people there. We did not know that the Chacaltaya Glacier was already in the last ten-thousandth of its 18,000-year life. No one did. It had shrunk dramatically – by about half – since the 1940s, but it was predicted to survive for at least another 50 years.

Twenty years later, in 2007, it was clear this had been optimistic. The lift had pulled its last skier uphill in 1998. Ice that still covered an area the size of several football fields in 2001 had since shrunk to two sad, separate patches. As an environmental object lesson it had turned into a miniature Aral Sea. Even then, those who knew Chacaltaya best said it would soldier on in some form until 2015 but two years later it was gone. There was nothing to see at the end of the track but schist. There was zip, as they say for emphasis out west. Nada. Zilch.

For a day or two the international media paid attention. Photographs were retrieved from archives and some of them bear a striking resemblance to the forlorn image of die-hard skiers on a strip of artificial snow at Davos, released by the SLF in 2016. The main difference between these images is that where Chacaltaya had a rope tow, Davos has a gleaming mountain railway. Each January this railway is used to lift the plutocrats attending the World Economic Forum up into the glittering mountainscape that overwhelmed and in the end seduced Hans Castorp. By this point in the season enough snow has usually fallen to impress a visitor – to "generate a winter feeling," as Christoph Marty put it. But what if, one of these years, the weather gods refuse to bless Davos even in mid-January? What if the whole place is bare and brown for the WEF? What will the plutocrats think then?

It's only a matter of time, and it's something I think about often. In fact I have a recurring dream in which a funicular of billionaires appalled to find Davos completely snowless resolves to spend what it takes to bring snow back from its glacial hideaways, down to the picture postcard villages where it belongs.

There are problems with this dream. One is it's a dream. Another is that unless appearances deceive, plutocrats tend to prefer golf to snow. Their reaction to snowlessness is unlikely to be one of horror. Until that changes, or world revolution puts climatological power in the hands of the people, snow will cling ever more poignantly to the world's high places. And instead of bringing it back down the mountains, humans will go on finding new ways to go up them.

CHAPTER TEN: BOUTIQUE MOUNTAIN ENGINEERING

"When I shall stand before God, the Eternal One will ask me, Did you see my Alps?"

Rabbi Samson Raphael Hirsch

The road from Krasnaya Polyana into the mountains used to be a goat track. It winds up out of the valley of the Mzympta River, which can be hot even in spring and autumn. The going gets cooler as it climbs through coniferous forest, and the gradient gets easier as it emerges onto the heathery upper slopes of the western Caucasus.

In February 2000 life changed for this idyllic corner of Russia: acting president Vladimir Putin arranged to be driven there in a motorcade of black Mercedes SUVs. Putin loves snow, especially as an expression of Russian identity. Where the track ended he clambered out in a red ski suit and declared the skiing in this part of his country to be comparable with the best in the Alps. It was his very first outing as a man of action. There was no hint of snow on the ground, which was not surprising. Sochi, where the Mzympta empties into the Black Sea, is a subtropical beach resort. Putin's compatriots did not yet know how seriously to take his boasts about Russia, but they soon would.

Since the collapse of Communism a few basic ski facilities had been built above Krasnaya Polyana. In the meantime Putin had tasted skiing in the west. He enjoyed it and considered himself a natural, but he disapproved of the oligarchs' colonisation of Courchevel, and he was smitten with the idea of building a Russian ski mecca to put Europe in its place. He busied himself first with intimidating his political rivals and crushing Chechnya's separatists; then he bid for the Winter Olympics.

Like Alexander Cushing 50 years earlier, Putin made exaggerated claims about the snow where he proposed to host the games. He brushed aside concerns about infrastructure, and won. His administration, realising it would have to build a large number of ski lifts in a small amount of time, turned to a privately-held engineering firm based in the quiet Austrian town of Wolfurt at the southern end of Lake Constance. Dopplemayr Garaventa was the obvious candidate. Putin spoke its language, literally, having learned German as a spy. The company was happy to be discreet about the scale of its Russian earnings and it had unmatched expertise in hoisting people to where the snow is.

For wealthy clients, Dopplemayr could even throw in a little style. Not long before the Sochi contracts went out to tender, the company had completed a four kilometre gondola from Whistler Mountain to Blackcomb in British Columbia with what was then the highest, longest unsupported cable span in the world. The Peak-to-Peak gondola rides on four track cables that each weigh 90 tonnes and took a year to wind. As its cabins pass over Fitzsimmons Creek they are 436 metres up, or one Eiffel Tower, or easily high enough to jump off with a parachute, which a base jumper did at the opening ceremony.*

Before that Dopplemayr had rebuilt the Galzig lift in St Anton as a fairground ride to save skiers the trouble of walking upstairs: each cabin, once loaded, is lifted onto its track cables by giant vertical discs as if onto a Ferris wheel.

The poker-faced engineers of Wolfurt seemed to have discovered in themselves a talent for showing off, and acquired a taste for it.

In 2012 they delivered the world's first *cabriobahn*, an open-topped double-decker cable car up the Stauserhorn in central Switzerland. By then the company had also undertaken to build the world's steepest funicular, in which passengers stay horizontal by sitting in self-levelling cylinders, and the outlandish two-stage $130 million Skyway Monte Bianco, whose high altitude wine cellar and 150-metre escape tunnel might have been designed expressly to appeal to the Blofeld Putin was becoming.

In the business of hauling people up mountains to snow, these flights of fancy are the knights in shining armour; the

* There is a story about the building of the Peak-to-Peak gondola that seems worth retelling even though it has nothing to do with snow. On their way to Whistler from the port of Vancouver in Washington state, the five 90-tonne spools of cable became stranded in a rail freight siding where the personal train of the chairman of Canadian National (CN) railways was blocking access to the unloading area. The chairman was in British Columbia on a golfing holiday. Herman Arns, a logistics chief hired to deliver the cables, invested a large sum from his employees' pension fund in CN railways, then phoned the company's investor relations hotline and demanded to be put through to the chairman as a stockholder. He was. The chairman's train was moved that afternoon.

drones servicing the queen bees; They are the loss leaders that may never pay for themselves but that lift us up, in the immortal words of Will Jennings, where we belong.

The starting price for a state-of-the-art cable car or mega-gondola in this second decade of the 21st century is around $50 million, or five times the cost of a world-class PGA golf course with clubhouse. One reason the Skyway cost more than double that is its length; it rises three-quarters of the way up Mont Blanc. Another is its fragile destination. It ends at a granite peak, the Pointe Helbronner, which is geologically similar to the Aiguille du Dru on the opposite side of the Mont Blanc massif. In 2005 part of the Dru known as the Bonatti Pillar, a granite spire the size of a respectable skyscraper, collapsed. It had been weakened by melting permafrost and simply fell off the main mountain into the valley below. A local geomorphologist, Ludovic Ravanel, described what happened in terms of failing glue: "The ice acts as cement in the flaws of the rock. But when temperatures increase and come near to 0°C, it does not play any more its cementing role."

The tension on cable car cables is huge and never lets up. Imagine a cruise ship pulling away from the dock having forgotten to cast off and instead of remedying the situation deciding to hold its course and keep its engines running for 50 or 100 years. The bollards would need to be securely fixed to the dock. In the same way the builders of the Skyway could not risk pulling the Pointe Helbronner off the face of Mont Blanc. The top station had to be anchored to the mountain without destroying it. Explosives were not an option because the freezing and thawing of ice in the rock had weakened it. Instead, a well was drilled like a giant root canal, 80 metres deep and eight metres wide, from the point itself into the bedrock below. The well was lined with concrete a metre thick and the top station's superstructure was anchored to that rather than the rock.

To get each of the four track cables up the mountain a Russian-built helicopter with twin contra-rotating rotors carried one end of an eight-millimetre pilot rope more than two kilometres long. That rope was spliced to a second, fatter pull-rope, and that to a third, and that to the full-strength cable, to be wound up the mountain at less than walking speed by winches which themselves had been disassembled to be flown to the top. Much of the winding had to be done in winter and it was so cold that the work crews had to stop every ten minutes to warm up.

The completed Skyway salved a certain amount of wounded pride. For 60 years the Italian side of Mont Blanc had lived, in engineering terms, in the shadow of the French. It had not been able to compete with the sheer daring and impertinence of the Aiguille du Midi cable car, which rose to 3,800 metres and in the early years of its construction had taken the moral high ground too: its historian, Denis Cardoso, claims the project offered safe work for Jewish labourers beyond the clutches of the Vichy regime during the Second World War. The Skyway has enraged environmentalists for intruding so brashly on the quiet splendour of the glaciers, but it has given Italy bragging rights, a new film location* and easy access to a spectacular domain of off-piste skiing.

---

* *Kingsman 2.*

Would it exist but for snow? It is hard to see how. The basic purpose of boutique mountain engineering is to bring snow within reach. Its absurd extravagance is justified by the safe bet that humans will remortgage to use it. These projects are sold by manufacturer to client and client to customer with snow front and centre in the CGI renderings and e-brochures. Snow is the prize. It is the one thing Vladimir Putin could not be sure of at Krasnaya Polyana, but his associates hired Dopplemayr to build 35 lifts anyway, in three years, including the world's longest, fastest "3S" cableway.

The "S" in 3S stands for *seil*, or rope. It's ropeway industry jargon for a gondola on steroids, in which each cabin runs on not one but three cables – one to pull it and two to support its weight. In the case of the 2.3-kilometre Olympia lift to the Krasnaya Polyana Olympic village, the client specified that it should be possible to replace the cabins with cars. Dopplemayr obliged.

The client was not so well served by the weather, although he had no reason to be surprised.

A fundamental fact of nature that accounts for much of the planet's snow is that the higher you go, the colder it gets. This is the adiabatic lapse rate in action – the rate at which air cools as it rises in the absence of any other added or subtracted energy, such as solar radiation or a blast of polar wind. This rate is about 5°C per vertical kilometre for moist air and 10°C for dry air. The difference is a result of the fact that in moist air the moisture condenses as it rises, giving off heat and slowing the process down. (Turning a liquid into a gas takes energy, and the reverse – including turning water vapour into clouds and precipitation

– releases it.) The underlying cooling is a result of Boyle's law, which states that as the pressure of a given volume of gas falls, so does its temperature. The upshot: if it's cardigan weather in Nice and cloudy in the mountains to the north, go there. It should be snowing.

Nice would be a reasonable place to host the Winter Olympics. In fact, defenders of the vote to award the 2014 games to Sochi point out that both cities are seaside resorts on roughly the same latitude. But if Nice were the host the snow events would be held at the nearest serious ski resort, which is above 2,000 metres. Thanks to the adiabatic lapse rate, average February temperatures at Isola 2000 are below freezing. Krasnaya Polyana, by contrast, is at 560 metres. Most skiing events at the Sochi games ended higher than that, but still a vertical kilometre lower than Isola 2000.

In an exceptionally cold year this need not have mattered. The western Caucasus can feel the effects of Russia's continental climate like the flick of a Siberian tiger's tail, but only rarely. In a normal year, February is springtime in Sochi. The prevailing winds that blow inland up the Mzympta valley are warm. They cool as they rise but they cannot be relied on to freeze until well above the spot where Putin declared the greatness of Russian skiing.

In 2014, February was warmer than usual. The Olympic hosts believed they had prepared for this scenario by hiring a snow consultant from Finland named Mikko Martikainen. In the end they might as well have hired Irving P. Krick or the snow dancers of the Western Shoshone Nation.

Martikainen did what he could. At his instruction, 16 million cubic feet of snow from the previous winter had been stored in shaded gullies near the start of the men's and women's downhill courses, high above the clammy valley floor. It was covered with thick isothermic blankets, which is not as odd as it may sound and in fact is a step up from the Swiss use of polythene. On the Stubai Glacier in the Austrian Tyrol, a snow protection team working with the University of Innsbruck claims to have saved ten metres of accumulated snow on a 10-hectare area at the foot of the glacier by covering it from May to October in five-millimetre white polyester fleece.

Martikainen also supervised a snowmaking system that purported to turn treated waste water into snow even at temperatures above zero. The technology does exist for this. It is more expensive and even more energy intensive than conventional snow making, which needs naturally freezing air. "Plus temperatures" snow starts melting as soon as it is made, but slush is sometimes better than nothing.

The system arrives on a convoy of trucks. It derives from industrial-scale cooling machines used in South African mines and costs about $2 million. Zermatt swears by it. Martikainen had it in reserve but opted to believe that temperatures at Krasnaya Polyana would fall and stay low. Instead, after a cool January, they rose and stayed high. This turned the snow on the lower slopes to "a creamy bisque", or so the sports columnist of the *Washington Post* complained. How could it be otherwise? Instead of freezing, daytime temperatures were between 10 and 18°C in the shade. In the sun they reached 30 on the middle Saturday, as they did in Palm Springs.

There was one more arrow in the snow whisperer's quiver:

salt. Salt lowers the freezing point of water, and not by a little. A 10-per-cent solution will not freeze until minus 6°C; a 20-per-cent solution stays liquid until minus 16. Spreading salt on snow may therefore seem foolish unless you want to melt it, but salt can, counter-intuitively, help make competitive sliding possible. There are some questionable theories on why this is so. One says salt melts surface snow to create a layer of water that freezes overnight into hard, fast ice. Another says salt gives body to the snowpack when warm temperatures are reducing it to slush. The real reason salt works, Professor Jim McElwaine of Durham University's Department of Earth Sciences suggests, is that it clears away warmer snow from a snowpack with many layers at different temperatures. If enough salt is applied to melt any snow warmer than, say, minus 5°C, this will run down into the snowpack and disappear. "The snow left on the surface will be minus 5°C or cooler and in a better state to survive the next day and provide good racing conditions."

There were difficulties concerning deployment of the salt weapon. One was that Sochi didn't have enough of it. "They had salt, but not the right kind," says Hans Pieren, a former Swiss ski racer who like Martikainen had been retained as a snow expert. "We already told them what kind of salt they needed the summer before, and they did not organise it."

Pieren had instructed the Sochi organising committee the previous September to order two tonnes of fine-grain salt, seven of medium and ten of coarse-grain Himalayan salt. The hosts had spent $50 billion on new roads, lifts, rinks, hotels and stadiums, a record for any Olympics, but they ignored the salt instruction. As spring temperatures turned to summer and TV cameras turned from the snowy peaks to the strips of slush on which athletes were expected to compete, Martikainen was wheeled out to reporters to reassure them all would be well.

"Don't worry about the amount of snow – that's guaranteed," he told the BBC, which took him at his word. He said that despite record temperatures for the ski jump competition "we've had very good snow – perfect, white compact snow. Absolutely no problems."

Behind the scenes, panic was setting in. An emergency meeting was called at the brand-new Park Inn Hotel, where Pieren's demand for ten tonnes of Himalayan salt was the main item of business. It was too late for the fine- or medium-grain salt to be of any use. Events staged on the lower reaches of the mountain were in danger of being cancelled. Only the coarse grain stood a chance of helping because, as Pieren told me later, the bigger crystals sink further into the snow, where they last longer and harden it to a greater depth. The small grains, he said, just sit on the top.

There was no Himalayan salt in Krasnaya Polyana, or down the new expressway in Sochi, or, it turned out, in the whole country. A snow drought was about to embarrass Russia almost as conspicuously as a surfeit had helped to save her in 1812 and 1943.

What was to be done? It was Pieren himself who made the call, to a friendly wholesaler in Basel. The Russian organising committee swung into action to divert a plane to Zurich to pick up the salt, and the following morning it was being sprinkled on the snow. The *New York Times* was first with the story of the great Sochi salt lift, and it was widely followed up. Whether the salt made much difference is another matter. It did not stop Shaun White, the world's top snowboarder, calling the half-pipe "mush". Lindsey Jacobelli, who fell in a snowboard cross heat, said it was like "landing in mashed potatoes". Andrew Weibrecht, who came second in the men's super-giant slalom, said everything "seems to slow down at the end".

Slush does that, and when the temperature is above zero at the dead of night slush is what you get. It is easy nowadays to click through historic weather data for almost anywhere on Earth, including Krasnaya Polyana. At no point in the middle six days of the 2014 Olympics did the air there freeze, even at midnight, 3am or 6am. I asked Hans Pieren if it was colder higher up, as the adiabatic lapse rate should have dictated. "It's not a question of altitude," he said. "It was warm everywhere, even at the top of the downhill. That was the problem. Nature did not make the snow there freeze."

The real tragedy of the Sochi games was that if Putin had spent more time making peace and less making war he could have hosted them 150 miles to the east on the slopes of Mount Elbrus. Running battles between Russian forces and local militants mean the British Foreign Office advises against "all but essential travel" to the region. But Elbrus is 800 metres higher than Mont Blanc and two miles higher than the peaks above Krasnaya Polyana. The temperature at the base of the mountain at breakfast time on the middle Saturday of the Olympics was a crisp minus 7°C.

Years before that, a few weeks after Putin's first public visit to Krasnaya Polyana, I paid a few roubles to ride up two old Russian cable cars to about 3,500 metres on Elbrus's western flank. From the top of the second, a drag lift disappeared into cloud. From the top of that an old snowcat offered to go higher still, from spring back into the middle of winter. I accepted and then skied down without stopping for about half an hour, first on powder, then wind-blown sastrugi, then perfect spring snow that never seemed to end. There was no cappuccino at the bottom nor any need for one. When basic meteorology is on your side, nothing else seems to matter.

CHAPTER ELEVEN: SNOWPOCALYPSE

"It takes a very big storm to fill a basketball hoop with snow."
Kevin Ambrose, *Washington Post*, Feburary 6th, 2018

In tepid England it's hard to imagine truly uncompromising snow, but planet Earth may be getting more of it. This is what I have secretly allowed myself to hope for the past 20 years, even though the only way to sustain the hope is to ignore sober long-term trends and focus on individual episodes. It may be unscientific, but I find it's good for the soul.

For instance, in November 1998 a mighty weather machine came to life over the Gulf of Alaska that was to earn a hallowed place in the annals of snowology. It even made it the subject of a cover story in *The Times*' science magazine, *Eureka*:

> A low-pressure system the size of Western Europe began circling over the vast wet wilderness between the Aleutian Islands and British Columbia. It sucked up moisture from the Pacific Ocean, pulled down freezing air from the Far North and churned them together like a monstrous invention from science fiction.

By the middle of the month the system… was a refrigerated vortex 1,200km across throwing off moisture-laden winter cyclones every couple of days. The jet stream blew these storms southeast towards land, where they chased each other down a shallow 500km groove between Vancouver Island and the Canadian coastal mountains. At the end of this groove, rising to nearly 3,300m squarely in the path of the weather, stands Mt Baker, a stubby white volcano that belches sulphurous fumes through an ice plug in its crater. In the lee of Mt Baker a two-lane road winds up to the local ski area, where it started to snow.

The snow did not stop for 35 days. After a break for Christmas it resumed, and by the end of December film crews and pro snowboarders had arrived from several continents like big-wave surfers drawn to Oahu after a hurricane. It kept falling for another six weeks. By the end of February 1999 the snowpack under Mt Baker's chairlifts was nearly 12m deep, despite constant clearing by snowcats and shovellers.

By the end of the season enough snow had fallen – supposing none of it had melted – to bury the whole of the Statue of Liberty except her torch. Roadside snow banks towered over everything that dared drive between them. Whole buildings were covered. Cliffs were reduced to ramps. Out-of-towners stayed away. There was simply too much snow to handle. But Amy Trowbridge remembers it as "the snowboarding winter of my life", and she has known a few. Now 39, she turned professional at 15 while growing up here, the snowiest place on Earth by record snowfall.

Oh, to have been Amy Trowbridge; to have known record-breaking super-abundant snow, to slide on, vanish into, emerge from spluttering and laughing and tell your grandchildren about. That was one possible response to the great Mount Baker winter of 1999 – envy. Another was curiosity. Was it an anomaly, conceivable only with a rare and perfect combination of latitude, altitude and weather? Or could it have been a signal? Could something like it happen again?

In America, of course it could, and for once I was in the right place at the right time.

The tremendous East Coast snowstorm that became known as Snowmageddon* arrived in Washington, DC on February 5th, 2010, a Friday. The previous evening the people who provide weather forecasts for the US federal government had abandoned all caution and started predicting snow with great precision. It would come the next morning at 10am, they said.

Schools closed. So did airports, stations, museums and government departments. Those who bothered to go to work on the Friday came home on the metro then went straight out in their cars to fill them with food as if there wouldn't be another chance for weeks. And there wasn't another chance, at least for days, because they stripped the shelves.

There had been time to see it coming. For a week, a storm system so big that satellite pictures could barely take it in had been churning across America. It had formed over the Pacific, dipped south for more moisture from the Gulf of Mexico and

---

* Also Snowpocalypse, Snowmygod, Snowcropolypse, Snoverkill and the Giant Clusterflake. Snowzilla appears to have been reserved for a storm of similar proportions that hit the US capital in 2016.

then headed north-east to merge with another system over Tennessee.

Between them, these two systems made a fat comma stretching from the Midwest to hundreds of miles south of Florida. There had been time to get used to the shape of the thing, time to leave town if you didn't like the look of it, time for the National Oceanic and Atmospheric Administration to measure its water content from above and below, time to model its route and work out with amazing precision when the moisture would be cold enough to give up flying and start falling. About ten o'clock.

In any snowstorm the first flakes are the best. From that point on there's always the fear that it will let up, but in those first seconds, when you look out or up and realise the flat white sky has started to disassemble, you can dream it will be big. And so it was. The snow that started that day at 10am did not let up for 36 hours. It deposited a record 32.4 inches (82 centimetres) at Dulles airport, nearly as much on our balcony and nearly three feet in parts of Virginia and Maryland. In a trick rarely performed by snow this far south, it filled basketball hoops to a foot above the rim. It cleared streets of moving vehicles and filled them up with snow. On the Saturday evening Dupont Circle was given over to a mass snowball fight organised on social media. Police, who could not escape by car, had no option but to take part. The rate of accumulation and the total depth for the storm were about half those witnessed by Douglas Powell in Big Whitney Meadow in 1969, but that just goes to show how lucky he was. It would not be fair on Snowmageddon to let it be diminished by comparison.

You went to bed with snow falling, woke up with it falling and went to bed again with it still falling. Old plans were cancelled, new ones made, and cancelled. Walking to Wisconsin Avenue could take a morning and there was no guarantee of finding a cup of coffee when you got there.

The skies cleared on the Sunday only to fill in again on the Monday with a second storm that dropped another foot.

Senator James Inhofe of Oklahoma and his family built an igloo in their garden and called it "Al Gore's new home". Stripped of politics and Inhofe's striking lack of curiosity (he is the most adamantine climate change denier in Congress, on the basis that humans can't change what God has wrought) the igloo made an important point. Snow records can be broken even when average air and sea surface temperatures are rising.

So what was going on? The NOAA offered a summary. The three key requirements for a big mid-Atlantic storm were all in place, it said: a high-pressure zone to the north funnelling cold air to the south; a low-pressure zone picking up moisture from the Gulf of Mexico before moving north from there; and a second low moving slowly up the Atlantic coast. The refrigeration and the moisture from multiple sources merged a little to the west of Chesapeake Bay and bore down on Washington, DC. There the storm ground almost to a halt and shed its load. Combined with the smaller storm that followed, the system "would become one of only three category five storms – the most extreme -- in the Northeast". Ever.

If Snowmageddon was only a function of the weather that might have been the end of the discussion. But what if it was a function of the climate too? There is a strong case that it was.

The case rests on Arctic ice, or it would if so much hadn't melted in recent summers. In 2012 a team of atmospheric physicists from Atlanta, Beijing and New York published a paper in *Proceedings of the National Academy of Sciences* (of the US). It was titled "Impact of declining Arctic sea ice on winter snowfall". It showed that between 1979 and 2010 there was a strong correlation between unusually small areas of sea ice surviving the summer and unusually large amounts of snow falling in the northern hemisphere in the late autumn and early winter. The paper put it in numbers: "a decrease of autumn Arctic sea ice of 1 million square kilometres corresponds to a significantly above-normal winter snow cover (more than 3 to 12 per cent) in large parts of the northern United States, northwestern and central Europe, and western and central China."

The team offered two explanations: "changes in atmospheric water vapour content over northern high latitudes" and "changes in atmospheric circulation linked to diminishing Arctic sea ice". The first category of changes are a simple tribute to our old friends Clausius and Clapeyron. Less sea ice means less light and heat reflected back into space, more light and heat absorbed by the dark waters of the Arctic Ocean, higher sea surface temperatures, higher air temperatures at sea level and thus *more moisture held in that air* to be distributed around the northern hemisphere by prevailing winds and pressure patterns.

The second category of changes relates to those winds and patterns. Less ice in the Arctic corresponds to higher-than-usual atmospheric pressure at sea level. This in turn seems to correspond to weaker-than-usual prevailing westerly winds around the top of the Earth. In colder times a solid low pressure over the North Pole could be relied on to preserve a strong polar vortex: cold westerly winds spinning anti-clockwise north of the 60th parallel. Weaker winds are a symptom of a weaker vortex, and they "meander". Like tourists heading south for the winter, winds released from polar vortex duty by rising pressure over an ice-free Arctic Ocean can stray thousands of miles from their usual stamping grounds. Having done so they can seem lost, shorn of momentum, circling slowly around and generally getting in the way.

In slightly more scientific language: "Weak westerly winds tend to enhance broader meanders that are likely to form blocking circulations. These blocking patterns favour more frequent incursions of cold air masses from the Arctic into mid- and low-latitude of western continents."

More moisture + cold air masses = looks like snow.

If the theory was right – if a warmer Arctic meant more snow further south – we were going to get a lot of it. And we did. In 2011–12 seasonal records were smashed in coastal Alaska. Rome recorded its deepest snow in 27 years. The English snow chronicler Fraser Wilkin reported that "for the first half of the winter the northern Alps were blasted by snowstorm after snowstorm and buried beneath a thick blanket of white". The next three winters weren't historic for northern-hemisphere snow but they all featured exceptional snow events. One was a record-breaking blizzard in eastern Massachusetts in January 2015 which Kevin Trenberth, a prominent climate change scientist at the US National Center for Atmospheric Research, attributed directly to global warming. Looking ahead, he said snowfall would start later and end earlier but midwinter snowfall would be heavier.

For snow addicts, it was a time of guilty consciences and secret hope. The idea that big snow might be linked to shrinking Arctic ice made enjoying it feel like handling stolen money. But still, big snow was big snow. One school of scientific snow thought was forecasting more snow at the start of the season; another in the middle. How bad, really, could that be?

From a strictly selfish point of view, looking ahead no further than the next few winters, the answer turned out to be not bad at all. The 2015–16 season was mediocre for much of the northern hemisphere but in January 2017 the first of a series of atmospheric rivers hosed the Californian Sierras with seven feet of snow. By the end of the season 44 feet had fallen on Mammoth Mountain and 56 on Squaw Valley – 21 feet more than Alex Cushing had boasted of as Squaw's seasonal average way back in 1955. In December 2017 the snowiest alpine winter in 30 years began with a wave of storms from the west. More followed over Christmas from the south and east. Above 2,000 metres snow lay three metres deep from France to the Dolomites for most of the winter. In mid-April the slopes above Engelberg in central Switzerland still had a snowpack five metres thick. In mid-May, when cowbells usually start clanging even above the treeline, villages in the Haute Savoie above 1800 metres got another 30 centimetres.

All this snow could have been accounted for by natural variability. Weather is fickle, after all, and "long-range forecast" remains an oxymoron even in the age of supercomputers. But if the snow of 2017–18 was part of a larger trend, did it mean more of the same or was there a catch?

There was a catch. Being virtually ice-free in the summer of

2017 was not the only remarkable thing about the Arctic in this period. It was alarmingly warm the following winter, even in the depths of the polar night when it should have been at its coldest. For nine consecutive days in February it was above freezing for at least part of the day at the Cape Marris Jessup weather station in Greenland, the most northerly land-based station on the planet. As Robert Rohde of the Berkeley Earth project told the *Guardian*, this was more time above 0°C for this station than for the four months from January to April for every year since 1981, combined.

Where had all the cold gone? It had gone south, to freeze the moisture to create the snow for a bumper bourgeois boreal winter. One NOAA scientist said if the Arctic was Earth's fridge, the door had been left open.

My children know what happens when the fridge is left open. It defrosts and pretty soon there's no cold anywhere. That idea, like death, is hard to think about without losing your bearings, which is why, aware of my cowardice and moral abdication, I prefer to think of the snowy present and recent past rather than of the uncertain future.

CHAPTER TWELVE: SNOWBUSINESS

"The fact is that the materials in any ski – about 2 kilos of hardwood, aluminium alloy, steel edges, fiberglass / Kevlar / carbon fiber and sintered polyethylene – cost no more than about $100–150. The rest is labor, debt service, margin and marketing hyperbole."

Seth Masia, President, International Skiing History Association

Snow costs nothing, but a view of it from a one-acre lot on Aspen's Red Mountain Road might cost $10 million. Snow falls silently but clearing it away can make the sound of a jet engine. Snow turns whole mountains into playgrounds but that's not good enough for a certain type of snowboarder, so a Swiss farm machinery maker has devised a 30-foot curving steel auger for carving half-pipes from piles of artificial snow. The starting price for the Zaugg Pipe Monster is $120,000, excluding the $350,000 snowcat on which it must be mounted.

Humans' relationship with snow is complicated and expensive. We make a tremendous fuss of it. I regularly remortgage my house in order to preserve access to snow and I don't stay on Red Mountain Road. One reason for the expense and complexity of this relationship is that it plays out along the edges of the world's snowy places, where snow is fickle. Another reason is fear. Cautionary tales about snow are etched in folklore and memorial stones, none more sobering than the Pioneer Monument in California's Donner Memorial State Park. It marks the spot near Donner Summit where a wagon train of emigrants from Illinois was stranded by snow 7,000 feet up in the Sierras in October 1846, and were not rescued for four months. The inscription explains that the height of the monument shows the depth of the snow covering the Emigrant Road on October 19th that year – a monstrous 22 feet.

"After futile efforts to cross the summit the party was compelled to encamp for the winter… Ninety people were in the party and forty-two perished, most of them from starvation and exposure." The inscription omits to mention that seven of the dead were probably eaten by survivors.

In the same place but a century later a 15-coach luxury streamliner express from Chicago to San Francisco was stranded by snow for four days with 226 passengers aboard. It was an awesomely overpowered train, pulled by three diesel electric locomotives with a total of about 60,000 horsepower. The Southern Pacific railroad controllers thought it would get through any blizzard but they were spectacularly wrong. By the morning of January 13th a storm that had started on the 11th had left 12-foot drifts beside the twin tracks over Donner Pass. It was still blowing. Even so, at 11.30, the City of San Francisco, for that was the train's name as well as its destination, headed west from the protection of train sheds at Norden, near the pass. By 12.15 it was stuck.

Norden is west of the pass, meaning the train was going downhill. Gravity was supposed to help it on its way, and so it did, pulling it deep into snow covering the tracks on an exposed bend just shy of Emigrant Gap. The snow continued that afternoon and night. By the morning there was hardly a gap between the train and the mountain. Spirits inside were high and meals were on the house, but rescue trains could not get through from east or west. A helicopter was sent up but couldn't land. Tracked Weazels were mobilised by the US 6th Army but defeated by the powder. The only outside help to reach the train in the first three days of its immobilisation were Southern Pacific workers who fought their way to it on foot, skiers with medical supplies and a doctor on a dog sled.

On day four the passengers walked off the train to a reopened road and were driven to safety. None of them died but an engineer did when an avalanche overturned his 50-tonne snowplough. A week after sliding to a halt, the City of San Francisco was eventually pulled from the snow one coach at a time by bulldozers, and snow was added to the list of elements to be mastered in the conquest of the west.

In Europe, mastery of snow has been achieved mainly by digging tunnels. The vertiginous topography of the mountains means the priority is to stop snow sliding onto humans from above. Since the Winter of Terror in 1951, 185 kilometres of road tunnels and 188 kilometres of rail tunnels have been bored under the Alps. These figures do not include the many avalanche sheds and tunnels shorter than five kilometres.

In North America, where the contours of the main mountain ranges are more relaxed, the priority for humans is to cover distance and clear snow out of the way. This has been accomplished mainly with machines. The Union Pacific railroad has taken over the line from Reno to Sacramento via Norden and has preserved three rotary blowers inherited from Southern, even though it only uses them once a decade. They were unleashed on the Donner Pass in January 2017 to clear the third heaviest snows on record in California, but railways are no longer where the glamour is in snow clearing; airports are.

In March 2003, 31 inches fell on Denver International Airport (DIA) in 24 hours, closing it for the first and only time in its history. It stayed closed for three days. Four thousand travellers were stranded. The weight of snow ripped one of the white tented roofs over the main terminal that are supposed to mirror the Rockies. It was the heaviest storm Denver had seen in 90 years – "a record-breaker, a backbreaker and a roof-breaker," said the mayor.

This was a costly humiliation, not to be repeated. Insurance claims reached nearly $100 million. Afterwards, the airport asked the Oshkosh Corporation of Wisconsin, the inventor of four-wheel drive technology in 1917 and a specialist in heavy-duty niche trucks ever since, to reinvent the snowplough, and the H-series all-wheel-drive blower was born. It had two engines. A 15-litre Caterpillar turbo diesel provided forward movement and a 16-litre, 650-horsepower turbo diesel powered the blower. The blower could hurl 5,000 tonnes of snow an hour a distance of 200 feet. In terms of work that equates to throwing a small car the length of an Olympic swimming pool each second.

Most impressive of all, the back wheels could steer along with the front ones, enabling the machine to advance crab-like down a snow bank, eating it without having to mount it.

The base price of the H-series was half a million dollars. Its slogan: Take Back the Runway. DIA placed a bulk order. When Oshkosh brought out the even mightier articulated XRS Extreme Runway System, which pulls a sweeper and an air blower as well as pushing a plough and snow blower, it stocked up on that as well at $640,000 a pop. The airport now has a 370-vehicle snow removal fleet including blades, brooms, blowers, ploughs, runway gritters, melters, loaders, bobcats and chemical tankers. It can clear a runway in under 13 minutes.

Denver airport is good at clearing snow, but not the best. Anchorage has never closed because of snow, and in more than 50 years Aomori airport at the north end of Japan's Honshu island has never caused a cancellation through failure to clear it.

Aomori's may be the snowiest airport in the world. Its average snowfall is 55 feet or nearly 17 metres. It has only one runway to clear but uses 38 machines to clear it. They are mainly Oshkoshes and Isuzus, but it's not the brand that counts. It's the teamwork and the timing. The Aomori crews drive in echelon at 20 kilometres an hour in an order honed through trial and error: plough, sweeper, plough, sweeper, three plough-sweeper hybrids, six more ploughs and then a blower. It takes them 40 minutes to clear the runway but they are careful not to start much longer than that before a scheduled landing if a blizzard is in progress. If they did the risk is that the runway would have filled in again before the plane arrived. Most of the drivers are rice farmers during the summer, but from November to April they are almost famous. They are members of a team known and copied up and down the snowy west side of Japan. They are "White Impulse".

In Frankfurt there is less scope for stardom because the new GPS-controlled snow clearing fleet from Mercedes Benz is driverless. There is an operator in the front plough but even he doesn't need to touch a steering wheel or joystick; only a keyboard. (None of this automation makes snow clearing in Frankfurt cheap. The airport spent 27 million euros on "winter services" in 2017–18.)

In Oslo and Helsinki (home of "world-class snowhow") the snow removal fleets are mainly Scandinavian-built. The airports are cleared with minimum fuss and maximum efficiency. In Moscow there is more of a sense of performance, and it is city-wide. The ridges of snow that build up along the edges of ploughed streets are eaten by mechanical bottom feeders with *zolotyi ruchki* – "golden arms" that pull snow onto conveyors. Conveyor to truck, truck to mixer, mixer to river: the whole life cycle of snow is accounted for.

In diplomatic compounds, teams of night sweepers see to it that snow hardly touches the ground. At Sheremetyevo airport the response is for the most part stoic and proportionate, although not always. I have been woken at 5am in the airport Novotel by the howl of a jet fixed to a flatbed truck. Subsequent enquiries suggested it was a Klimov VK-1 engine, a knock-off of the Rolls Royce Nene, used in Mig 25 fighters in the Korean War. They develop 6,500 pounds of thrust at up to 750°C, which makes them useful for clearing snow and ice from large areas of concrete, especially when jet fuel is as abundant as water and no one minds about waking people in the Novotel.

I bear no grudge. The Russians have reason to thank snow for its services to the motherland, but also to resent it for making life in the Gulag even harder than it already was. "How is a road beaten down through the virgin snow?" Varlam Shalamov asks in the first of his *Kolyma Tales*. "One person walks ahead, sweating, swearing, and barely moving his feet. He keeps getting stuck in the loose, deep snow…" Five or six follow in line abreast to stamp out a road suitable for tractors, but "the first man has the hardest task, and when he is exhausted, another man from the group of five takes his place".

No wonder a few Russians went crazy when history eventually let them. They kept their connection to snow, but dropped the deference. The Sochi Olympics were one expression of the new mood; Mikhail Prokhorov's sex parties at the Hotel Byblos in Courchevel were another.

It was within two years of the end of the Soviet Union that the first $1,500 bottles of Bordeaux started appearing on the menus of modest pizzerias in Courchevel. I know this because after working the 1992 season there I returned to help my replacement prepare for 1993. Modest pizzerias were all we could afford. That was the summer that the menus were reprinted in Cyrillic. That was the year, for the oligarchy, that Courchevel took off.

Prokhorov is six foot eight and fond of living large. When he was 30 and a banker in Moscow he went into business with the man who devised the scheme that created the oligarchs in the first place. That man was Vladimir Potanin. The scheme was to lend the Kremlin administration of Boris Yeltsin enough money to stay afloat, in return for stakes in the country's state-owned industrial giants and utilities. Potanin had his eye on Norilsk Nickel, a mining conglomerate 175 miles north of the Arctic Circle. There, in a catacomb of tunnels a mile below the permafrost, Stalin's slaves dug for palladium and platinum as well as nickel, and found them in prodigious quantities.

The snow that falls on Norilsk is a travesty of snow. The flakes form around industrial soot. They fall through clouds of sulphur dioxide and tiny droplets of smelted nickel, and land as a ready-soiled blanket.

The snow at Courchevel is brilliant white and groomed to perfection, and young Prokhorov fell hard for it. From 2001 to 2007 he worked diligently by oligarchic standards, shuttling between Moscow and Norilsk even in the depths of the long Arctic night, cutting his company's workforce by half, boosting its share price by a factor of 27 and raising his personal net worth to about $7 billion. He relaxed in Courchevel every Russian New Year, and in 2007 he flew in seven attractive young women to relax with him. He was arrested on suspicion of filling the most expensive suites of the Byblos with prostitutes. He and his lawyers argued that the women were in fact models, and after 88 hours in a cell in Lyon he was released. In that time journalists descended on the resort to ask what sort of company the tall Siberian nickel miner kept. "These girls, you see them all the time, they never ski, they walk around Courchevel on high heels," one visitor told Reuters. And it's still true. You do. Prokhorov reinvented snowbusiness as St Tropez at 7,000 feet.

It helped that he could fly almost to his hotel door. At the time of the Prokhorov affair the press pack picked up on the idea that the oligarchs liked Courchevel because it has its own airport at 2,000 metres. He could arrive in his own jet, they wrote. This wasn't true. Courchevel does have an "altiport" but its steeply sloping runway is only 500 metres long and jets are not allowed to use it. Nice airport, though, is 45 minutes away by helicopter. That will be about 16,000 euros each way in a spacious six-seat Eurocopter, but the point is that an oligarch need never be more than an hour from his yacht.

On landing at the altiport, the walk to the Range Rover need not be more than a few steps. Since the unpleasantness of 2007 a large number of multi-storey mansions in the savoyard style have been completed on a gated compound off the Rue de la Vizelle, minimising the risk of disturbance by gendarmes. These palaces of private fun rent for up to half a million euros a week in high season, which is no more than you might be paying for the yacht. *Service hôtelier* is available for those who choose not to pre-position their own staff. Snowmobiles, snow kayaks, flightseeing, accompanied parapente descents and of course the usual suite of indoor pamperings are available for non-skiers. Private instructors at 450 euros each per day serve as tickets to the front of every lift queue. Candidates for the most expensive skis ever sold are those created specially for Courchevel in 2008 by the French firm, Lacroix. Only ten pairs were delivered to the company's boutique there, each in a leather travel trunk with carbon fibre poles, gloves, goggles, bindings and a full season lift pass. The price: 50,000 euros per trunk.

Groups of mixed ability and none can meet for lunch at Nammos; a four-by-four can get you close enough to walk there in flip-flops. If everyone is having fun, lift-off can be delayed till dusk.

Such are the rough contours of money-no-object snow in the early 21st century. It has changed beyond recognition since the 1860s, when Johannes Badrutt started inviting English summer guests at his Palace Hotel in St Moritz to return in winter with the promise of a refund if they did not enjoy themselves tobogganing. It has changed almost as radically since the 1930s. Then, Averell Harriman paid stars to try out his own private Idaho in Sun Valley. They were at least expected to pose with their instructors for photographs on the summit of Bald Mountain. Nothing is expected of the oligarchy and its entourage except maximum bling. Nothing is expected by them except instant and complete gratification.

Mass-market snowbusiness has changed more subtly. For about 15 years the total number of ski trips taken by humans each year has been stable at roughly 400 million, according to the tireless Laurent Vanat, who publishes the annual *International Report on Snow and Mountain Tourism*. That number includes 54 million trips taken by Americans (three times as many as by spectators paying to see NFL football games). They spend about $5 billion a year on ski gear alone. The 400 million number keeps more than 22,000 ski lifts busy in 67 countries including Israel, Algeria and Lesotho. It includes trips taken by 1.5 billion Brits a year at a cost of nearly $3 billion (more than they spend on boating) and it is ticking up as China falls for skiing: 57 new resorts were opened there for the 2017–18 season alone. Here and elsewhere, skiers have been persuaded to accept fake snow as the price of climate change. In return, sliding down snow has been made easier.

The first of these changes is a triumph of technology and

marketing, even if it is an aesthetic and environmental disaster. "People do not care about the snow, they care about the sun," says a senior snow-making executive in the Dolomites, where ultra-modern wide-bore snow cannons with smooth aluminium cowlings line most ski runs because so much more of Europe's snow comes from the north-west than from the south and the Dolomites lie on the wrong side of the watershed.

This executive, quoted by *The Economist*, is surely lying. People do care about the snow. They yearn for its softness, its lightness, its slipperiness, its differentness and the way it falls naturally from the sky. That is why they will pay a month's wages for a weekend in its presence. If what they find is fake they will accept it as a surfer accepts an artificial wave machine, but that doesn't mean they do not care. Why else would ski resorts try to euphemise fake snow out of existence?

In Austria it is "technical" snow. In France it's cultured. In Italy it's programmed. North American resorts announce the start of "snow making" when temperatures fall low enough, as if it were as traditional as Thanksgiving. It has become traditional but there is something desperate about the rasp of the machines as they labour through the night to build piles of ice crystals big enough to bulldoze into stripes across the mountainsides.

All this is done out of necessity, not joy. More than 200 resorts in the eastern Alps will depend on fake snow for their survival if average temperatures rise by another two degrees. Every new resort being built in China will depend on it come what may. Even now every World Cup ski race is run on artificial snow because it comes at the touch of a button, packs down hard and gives each racer roughly the same ride. They like it for being fast and fair. Otherwise the best that can be said of fake snow is that it makes real snow seem even more sublime.

Real snow means business too, although it is a flakier proposition. It has an exceptional ability to separate people from their money (some Chinese devotees call it white opium), but the challenge is to be ready for it when it comes, with new spending opportunities in case old ones begin to pall. This was already true in the 1970s. The upshot then was the snowboard, which allowed baby boomers to look different from their parents on the mountains, to "ride" instead of slide, and to claim a cultural right to fill gondolas with the rich perfume of hashish. Later the snowboard would create a new breed of star on a new sort of snow creation – the half-pipe, *raison d'être* of the Zaugg Pipe Monster. In the meantime the novelist Amy Tan and her husband, Lou, went skiing in Squaw Valley with their teacher, Seth Masia.

The year was 1993. Amy was a keen skier, but not a good one. She had been raised in San Francisco by her Chinese parents before moving as a teenager to Switzerland. She went there with her mother to escape a curse – her father and a brother had died of brain tumours in the same year – and they settled in Montreux with little foreknowledge of the sort of people they would meet. At her new school these people turned out to be boys who smoked Gauloises and girls who "wore lynx coats with nothing underneath". They skied locally in PE lessons and on weekends in Gstaad, where, on the first run of her life, Amy fell to avoid colliding with the Queen of Sweden.

"I was the girl who couldn't run a relay race without falling down and throwing up," she would write years later in *Ski Magazine*. "I was the player who sprained her finger just looking at a volleyball. I was the bungler designated to stand out in right field where baseballs were seldom hit. The one time a girl really did hit a fly ball out there I was jeered at for running away from the ball."

Even so, she found that she loved skiing. Back in America she stuck at it year in, year out, "despite bad equipment, ludicrous ski outfits and humiliating face-plants". When I phoned her to ask why, she said: "It's the only sport I do. It comes with danger, beauty and this exhilaration if you come out of it alive."

She met Masia, who is a ski intellectual of the first order, at a writer's workshop in Squaw Valley in 1985. They became friends. She became famous with the publication of *The Joy Luck Club* in 1989. She did not let that interfere with her skiing; nor did it make her any better at it. In 1993 she and her husband were still intermediate skiers, Masia remembers. "Squaw has a reputation for very steep and difficult snow conditions because the snow comes off the ocean wet. Amy and Lou were brave, but they didn't handle it in great style. Lou is a powerfully built guy. He tended simply to use his strength to muscle the skis around. Amy is slight and wasn't able to do that. She depended much more on balance."

Masia was in charge of testing new skis for *Ski Magazine*. This meant he was one of the few people in the world who knew that a young Slovenian engineer named Jurij Franko had single-handedly redesigned the ski in a nondescript factory at the foot of the Julian Alps. Franko's Elan SCX skis were much shorter than traditional ones, and fatter at each end, but still thin under the boot. SCX stood for side-cut extreme. Whereas traditional skis were long and straight and counter-intuitive to turn, these had a waist and turned themselves. All you had to do was tilt them just enough to set the edges on the snow and "carve" (the word is everywhere now; it wasn't then). The skis would then describe an arc of which the edge was a part. In the case of the Elan SCX this arc, provided the skier didn't skid, would be part of a circle with a radius of 15 metres – about a third of the turning radius of traditional skis. The SCX was designed as a giant-slalom ski, and to be able to ski a 15-metre radius turn just by setting the edges and letting the skis run was a giant-slalom revolution. The new skis, Masia wrote, were "blazingly fast". The first time they were used in races, in Slovenia, they won eight of the top ten places. Carving was born.

By 1993 Masia had been sent test pairs of SCXs and a similar ski from Kneissl called the Ergo. That day in April at the tail end of the season he invited his slightly built celebrity novelist pupil to try them.

"It was like gliding," she says.

Her teacher wrote a more detailed report: "She was immediately able to carve clean turns in spring corn over rotten crust. I put her husband on the SCX and he could do the same... They may have the honour of being the first ski-school clients ever to learn to carve on modern shaped skis."

Almost no one who tried shaped skis went back to old straight-sided ones, just as almost no one who tried a big Prince tennis racquet in the 1970s went back to a Dunlop Fort. The new skis were so short that they put an end to the machismo that attached to long ones. "There was some resistance from very good skiers but that tended to melt away as soon as they tried them," Masia says. "Three years later, you couldn't race without them."

That is only a mild exaggeration. What happened three years later was that a young skier from northern New Hampshire was handed a pair of radically waisted skis made at the K2 factory in Puget Sound, a short ferry ride from Seattle. These skis had a side-cut depth of 14 millimetres – they were 14 millimetres narrower at the waist than their average width – and a turning radius of 22 metres. They were called K2 Fours and their first outing in competition was at the 1996 US Junior Championships at Sugarloaf, in Maine. The skier who showed what they could do was a rebel with a cause, which was to win. He had skill, strength and recklessness in abundance.

His name was Bode Miller. His style was his own and it suited the new skis. It was about them, not him. He allowed himself to lean back when generations of coaches would have screamed at him to throw his weight forward. He could use muscle and a sixth sense for the course to recover from situations that should have ended in the netting. He didn't care how he looked as long as his skis were taking the fastest possible line down the slope. He won three of the four races he entered that year at Sugarloaf, and the die was cast. Overnight, Masia wrote, every racer in the country needed a pair of K2 Fours just to be in the game.

In my ski racers' hall of fame Miller would have a place, but he would have to share it with at least a dozen men and women (16, in fact) who have more World Cup victories to their names. He is not the greatest skier of all time. That would be Lindsey Vonn or Ingemar Stenmark, probably in that order.* He is not the greatest downhiller of all time. By any measure –

* Stenmark has won more races than Vonn, but all in slalom and giant slalom. Vonn has won in all four alpine disciplines.

medals, podiums, excitement, name recognition – that has to be Franz Klammer. He is not an extreme skier, a free-rider or a slope-styler, although he could be if the spirit moved him. He is, however, a pioneer. He reinvented skiing and in the process earned a place on the hard road to victory in the longest, fastest ski race of them all.

WENGEN

CHAPTER THIRTEEN: SNOWMADS

"Snow couloirs may be considered as natural highways placed, by a kind of Providence, in convenient situations for getting over places which would otherwise be inaccessible. They are a joy to the mountaineer... but they are grief to novices."

Edward Whymper, *Scrambles Amongst the Alps*

When the great journalist and raconteur George Plimpton died in New York in 2003, his obituaries gave me an idea. Plimpton was famous for his prose, but probably more famous for three rounds he once dared to box with the legendary welterweight Sugar Ray Leonard. For that and similar stunts he was remembered as the founder of "participatory journalism".

Doing stupid things to write about them is a time-honoured way of filling newspapers. Plimpton had gone one better. He had turned this into a way of making his dreams come true, and he had given it a name. My idea was simple: go skiing in his honour. I would try to get someone else to pay me to undertake a piece of participatory journalism on the longest, fastest downhill on the World Cup circuit.

There was no evidence that Plimpton had any interest in

snow or skiing but that wasn't the point. His death was an opportunity. I had a hunch that his ghost would approve if I could pull this off. The downhill in question was the Lauberhorn, above Wengen in the Bernese Oberland. I would ski it as fast as a deluded amateur could be expected to, after it had been groomed for competition by the Swiss army and before it had been cut to pieces by professionals. Special permission would be needed. The window of opportunity would be narrow. The risk of life-changing injury would be high. That is what Plimpton would have claimed, so I did too.

My editor at the time wasn't much interested in editing or budgets. He heard me out and nodded. It took a while to realise he was saying yes.

The Lauberhorn is Switzerland's answer to the Hanenkamm, which Austria claims is the most dangerous downhill in the world. The truth is more racers fall on the Hanenkamm, above Kitzbühel, because it's so steep. But more racers have been critically injured on the Lauberhorn because it's faster and more tiring. The fastest speed ever travelled on a World Cup downhill was 100.61mph by the Frenchman Johan Clarey on the Lauberhorn in 2013. Its average gradient is 25.3 per cent, or one in four.

It has two jumps near the top. The first is the Russisprung, a few seconds below the start gate. The second is the Hundschopf, the dog's head, the most dramatic jump in ski racing. From above, the Hundschopf is framed by two dark cliffs; the course goes between them. From below, skiers seem to take off as if launched by a catapult and to fly through the air for the hell of it. In fact they're desperate to cut their time in the air because on landing they have only a fraction of a second to absorb a shallow compression and position themselves for a right-hand turn against the camber of the mountain. This compression is now named after Josef Minsch, who was helicoptered away from it in 1965 and spent the next ten weeks in hospital.

The Lauberhorn is 2.78 miles long – nearly a mile and a minute longer than any other course. "It's like four downhills all kind of stuck together into one," Miller has said. "You can't see well because you've got no blood going to your head. All the blood in your body's going to your big muscle groups to keep you standing up. Things go grey. You start to get tunnel vision. And then you have two of the most aggressive technical turns in all of downhill skiing."

The first of these turns is a right-hander ending in a hard left and a jump. The finish line is a few metres beyond the jump. Miller says that by the time he took this jump in 2007 he was nearly blacking out. He fell on landing and crossed the line in a flat spin, but still won the race by a second and a half. He won the following year as well.

In my teens the Lauberhorn became the primary focus of my snow addiction. When I left school all I wanted to do was stand in a snowstorm on the edge of the course and watch the world's best skiers fly by. So with a long-suffering friend I hitched from Calais to Wengen and tried to sell glühwein prepared on a camping stove on the side of the slope. That didn't work, but we did see the race (although "see" is the wrong word; we heard the clattering that downhillers' skis make as they vibrate on snow injected with water to turn it into ice, and we caught the briefest glimpses of men with names like Weirather and Wirnsberger hunched over in their fight with gravity and lactic acid). Race over, we hitched back to England.

Nineteen years later I needed a guide who could persuade the authorities. I secured the services of one whose regular clients included John le Carré. His name was Paul and he knew the people who prepared the course. They said they would turn a blind eye as long as we were off the mountain before their working day began, and this meant skiing the course while it was still getting light.

We were on the train to the top before dawn. As the north face of the Eiger emerged from the dark a mile above us we were ready to go. A battalion of Swiss mountain infantry from Fribourg had been compacting the snow for a week. It fell away from the start gate smoothed and pummelled and shaped like the back of a polar bear. "If you want to go speedy, this is the place," Paul said. "You get to maximum in two, maybe three seconds." And he was off.

Getting to 90mph in three seconds involves twice the maximum acceleration of a Porsche 911, and even getting to half this on skis is a noisy business. I don't know about racers' helmets but with cheap rented ones the air howls in your ears. As the howl rose to a scream Paul glanced back and dropped out of sight. I tried to keep up but he was going fast by his standards, not mine. I blinked at the Russisprung and bottled the Hundschopf. On the easiest part of the course below the railway bridge, I took a breather. In the same place, in 1997, a junior racer blew through two layers of fencing and broke both legs on a tree.

On the Hannegschuss, the fastest section of the course, I had visions of my right leg declaring sudden independence. I stood up and turned across the slope. I chickened. We went back up and had another go. I chickened again.

Each time an argument in my head ruled out any chance of focusing on the job. It was an argument between lost youth and grudging wisdom, between valour and discretion. Discretion won, so I'm alive and able to confirm that World Cup downhill skiing has something in common with chess. Those who do it for a living operate in a class and a world of their own. The difference is that chess masters risk nothing but pride and foregone prize money. Downhillers risk everything.

By our second run the sun was up and the Fribourg infantry were on the mountain for one last day of scraping and smoothing before the race. Not far below the Hannegschuss they were putting up hundreds of metres of special blue netting with tiny holes along both sides of the Lauberhorn's final S. The size of the holes is an important and relatively new development.

When a promising young Austrian called Gernot Reinstadler reached this point on January 18th, 1991, the only netting was standard issue and the holes in it were large. Reinstadler was exhausted. Remember Miller: *"You can't see well because you've got no blood going to your head. All the blood in your body's going to your big muscle groups to keep you standing up. Things go grey. You start to get tunnel vision"*. As Reinstadler tried to set himself up for the very last turn his right tip caught in one of those big holes. The ski should have detached at once but didn't. Here was a 21-year-old with a career ahead of him and bindings cranked tight for the race of his life. His injuries were too horrifying to describe. His leg was almost pulled off then and there. As he slid to a stop the snow was streaked with blood. People who were there said they could not bring themselves to look. He died ten hours later. The doctors said massive internal injuries were the cause, but a memorial at the bottom of the Lauberhorn says it was his love of skiing.

Skiing for the love of it is easy to explain. Skiing for speed is harder. It's pure but futile, beautiful but dangerous and undertaken at the outer limits of human strength and sanity.

Robert Redford tried as hard as anyone to show what it takes to stay upright at 90mph, in *Downhill Racer*, which was filmed on the Lauberhorn. He was also fascinated by the idea of the rough-hewn American interloper taking on a European elite at its own game. Life imitated art with Miller, and before that with wild Bill Johnson, who won the race in 1984. In between came the Canadians: Ken Read, Dave Irwin, Todd Brooker and Steve Podborski, crazy Canucks in bright-yellow suits, winning when they stayed on their feet, cartwheeling into horror crash compilations when they didn't.

They were a new sort of nomad. They were busy in winter, like the reindeer followers of 10,000 BC, but they moved in minibuses and slept in cheap hotels. In principle they followed the snow. In practice they followed sponsors' timetables and tradition. It was a strange, obsessive hunt for new ways of losing nanoseconds, with endless distractions and minimal chances on a given day of ending on the podium.

At 90mph anything could go wrong. Bindings cranked on

maximum DIN* settings could release when you needed them most. In 2005 Miller lost a ski 26 seconds into the Bormio downhill in Italy and kept on going on one leg, waving the other in the air as if listening to *Swan Lake*. When Todd Brooker lost one on the Hanenkam in 1987 he fell head first into a gate, knocked himself out and span on down the course like a rag doll for perhaps 200 metres. His recovery took four months.

The professional ski circus has expanded but not changed much since then. Tribes form around each new style of competition and their members might have hoped for more freedom to follow where the snow gods lead. This doesn't seem to happen. Like the hard men and women of the World Cup circuit, the reigning hipsters of free-ride and slopestyle go where their sponsors tell them to.

The day after Dave Rosenbarger died on the Pointe Helbronner in 2015, the Freeride World Tour staged its annual Chamonix contest on the Pente de l'Hôtel on the north side of the valley. In free-ride competition skiers can pick their own line and are judged for style and daring as well as speed. The Pente de l'Hôtel is a 40-degree mixed slope of cliffs and couloirs. It's high and steep but faces south. Snow that falls on it melts fast and the cover was thin. Rocks were exposed under many of the jumps the competitors were expected to incorporate into their runs. Some of the skiers had known Dave but they couldn't miss the event to mourn him. They all had careers and social media profiles to consider. One seriously viral post can be life-changing, in a good way.

---

* DIN stands for Deutsches Institut für Normung. Translation: tightness.

Sam Smoothy, from New Zealand, had chosen free-riding over traditional alpine competition. He had the name, the build and the attitude to be a star, but wasn't one yet. If getting there meant pushing his luck, his thinking was so be it. He'd jumped off a 30-metre cliff in Verbier in 2011 and survived, which seemed to have affected his approach to risk. "Six foot four and bulletproof" is what he said to his girlfriend and himself before each run, and those were his parting words before setting off on the half-hour hike to the top.

I watched with relief from below as he seemed to bottle the descent. He explained afterwards that he had felt a cable in his left boot snap in the starting gate. Too much flex in the boot on a hard landing could have blown his knee. Too much flex also ruled out hard right-hand turns and he needed to make several of them to hit the last jump on the run at exactly the right speed. Without total control, he said, "you either go pussy or you go too big".

Why bother at all? Taking huge risks for a few seconds of footage of a jump off a ledge into deep snow may seem ridiculous, but I don't think it is. I think it's art. Consider: it's self-expression. It goes beyond eating, sleeping and procreation. It's something our species does that others on the whole do not, and it produces an artefact as surely as Monet did when he headed out with paint and brushes into the snow near Giverny.

None of which alters the fact that hitting a jump at the wrong speed can be life-changing, in a bad way.

I had watched Smoothy's run sitting on the snow next to a woman introduced to me as the toughest person on skis of either gender, ever. Her name was Marja Persson. She was from Sweden and had been skiing since she was two, with a break of two years that started one February day in 2011. The Freeride World Tour had taken her to Kirkwood, north of Lake Tahoe, where she attempted a jump on a slope she had been allowed to look at but not ski down in advance.

"I came in a little too hot and saw I was going to land on rock," she told me. "I tried to overshoot but ended up hitting the rock right on my ass. The rock there is volcanic and sharp… It wasn't so nice."

Persson thought she had broken her femur but it was much worse than that. She had suffered extensive internal organ damage, while a shattered pelvis and broken lower vertebrae had left her spinal cord completely unsupported. Local regulations forbad a helicopter rescue and the use of morphine on the mountain. It was, she said, the worst pain she'd ever felt. It was followed by three years of hospital, surgery and rehab and it was four years before she could ski again without pain.

There is a safer sort of snowmadism. It doesn't pay and it isn't glamorous but it does mean following the snow as closely as the Gwich'in Nation follows the Porcupine caribou herd from Yukon to Alaska. I know of only one true practitioner, a French Canadian named Evans Parent. For 13 years, for three or four months a year, he has sought out the best powder he can find and skied it. He goes with friends or his father when possible. Otherwise he migrates alone. "The goal is simple," he wrote when starting out. "Ski as much powder as we can, spend as little money as we can and have a good time."

Parent suffers for his art. His standard procedure is to convert cheap second-hand cars into camper vans and skin* up mountains instead of buying lift passes. In the British Columbian backcountry, a favourite destination where there are often no lifts anyway, he thinks nothing of walking up the same mountainside several times a day if the snow is worth it. He is not an absolutist – he'll sleep on a sofa if offered and buy a ticket if he has to – but he is a purist. Those who join him for sections of his annual ski safari expect to sleep rough, sweat a lot and eat off camping stoves. Every turn he makes is in the old, fluid telemark style, kneeling before nature. Every turn that crosses someone else's tracks is considered sub-optimal.

He keeps a retro, unselfconscious blog with minimal commentary, partly because it's in English and English is his second language. When he started it he explained that his winter generally begins with a road trip from Quebec to the Rockies with "no sleep until we can see the moonlight reflected by the snowy peak". The next day's task is to find "a little piece of paradise where the snow falls so fast that if you stop moving it covers you or gets into your sandwich".

Parent's version of a tough choice is one between heading straight across Canada to the Selkirks west of Calgary or bearing left for Utah. His idea of bliss (having chosen Utah) is to be trapped for the night in Little Cottonwood Canyon with the road up from Salt Lake City closed by a blizzard.

For the first few years the safari stayed in North America. In 2008 it went international – first to Japan and eventually to Gulmarg in the Himalayas; Karakol and Arslanbob in

* Walk with skins stuck to the soles of his skis to stop them sliding backwards.

Kyrgyzstan; to Greece, Norway and Argentina. The full list is contained in a word cloud on the blog that serves as a human counterpoint to the pixel maps of the Rutgers Global Snow Report. If you want to know where a snow addict who has structured his life around his addiction goes to feed it, it's there at a glance. Everywhere he's been is in the cloud, together with the number of related posts. This, like the font size, is a proxy for the number of visits made.

Greece, Hungary, Montana and Yukon each get one post.

Argentina, Bulgaria, Quebec, Washington state and Wyoming each get two.

Austria, Kazakhstan, California and Chile: 3

France, Georgia, Kyrgyzstan and Italy: 5

Gulmarg, India, Switzerland: 6

Alaska, Norway, Utah: 8

Hokkaido: 23

British Columbia: 31

Japan: 35

Interesting omissions include Colorado, New England and Russia, but still, this is an enviable list. Occasionally I've wondered if mine could qualify me as any sort of snowmad. I doubt it. It has been augmented by odd pieces of serendipity – the lightest snow I've ever known was in Aspen, where I'd been sent to cover a Monty Python reunion.

The deepest was in Utah (an execution one year; a Robert Redford interview the next). The heaviest was on Mount Hood in Oregon, after a murder trial in Portland, where the jury handily reached a verdict early on a Friday afternoon. But this is random stuff compared with Evans Parent, who is so committed to snow chasing that he can be hard to run to ground. The first time I tried was in 2012. He would reply to emails but after long delays, with messages like "I was out skiing so I missed what could have been the Skype appointment". Or: "I am actually in Vancouver and heading to Washington state tomorrow. There is a huge system bringing tons of snow."

He once let slip that he'd be having a 90-minute layover in Heathrow en route to Kyrgyzstan, but I couldn't be there. As a consequence he's taken on a semi-mythic status in my imagination. He is the complete self-actualisation of the snow addict, more avatar than human; the Odysseus of winter.

CHAPTER FOURTEEN: THE FUTURE OF SNOW

"For the Lord spake unto Job:- Hast thou entered into the treasures of the snow?"

The Book of Job

Humans' relationship with snow has been long, complicated and exhilarating. For at least 35,000 years snow has been a marker of our seasons and a store of our water. For as many as 10,000 it has been a go-anywhere winter transport network for anyone with the wit to fashion a pair of runners and the strength to tolerate the cold. For as long as anyone can remember it has been a cause of complaint, but then we're given to complaining. Snow has been a source of wild and outlandish thrills. It's true that polar bears use their bodies as toboggans when that is easier than walking, but *Homo sapiens* deserves some credit. In the business of sliding on snow no other species comes close.

The Zaugg Pipe Monster may sound comical and look like a giant metal insect but it exists and it works. You could argue that there would be no front-side double-cork twelve-sixties*

* A jump only performed by Shaun White, history's greatest half-pipe snowboarder.

without it, and you could state with confidence that life would be duller without them. The 120-metre ski jump – the "big hill" that sidles onto Eurosport each New Year – may fill any normal soul with fear, but a highly evolved sub-species of human has conquered that fear and uses the big hill to fly. Gernot Reinstadler may have died on the Lauberhorn, but the course is still celebrated as a great natural theatre of excitement and the people of Wengen, at the foot of the mountain, celebrate snow whenever it falls. They know something that the rest of us are only beginning to accept.

We have taken snow for granted for too long. It teases us with sporadic reminders of how deep and crisp and even it could be, but on the whole it's making itself scarce. It's skulking in the high country like the yeti it has occasionally betrayed.

The question is whether snow's retreat is irreversible, and it may be. It is possible that children born in England this century will never see a white Christmas, except perhaps in the Peak District and on the Lakeland fells. Already, children born this century in Moscow are used to rain in winter even as their parents struggle to adjust to it. Muscovites who knew the Soviet Union also knew deep-frozen winters that seemed unchanged since Anna Karenina and Count Vronsky glided along white streets in sleighs and furs, as enraptured by the snow as by each other.

But it's too soon to declare snow doomed. Even if there was no alternative to sitting back and watching it melt, that would be too miserable to contemplate. So what are the alternatives? What would it take to turn the climatological clock back to the winters that inspired Tolstoy on one side of the world and

Laura Ingalls Wilder on the other? Where in the world will it stay snowy come what may? And is it possible – in practice, not just in theory – that some of these places might get more snow in a warming world rather than less?

I have a picture on my phone that cheers me up whenever I look at it. It's of an eight-year-old in helmet, jacket and ski trousers, sitting on a stone wall, waiting for a bus. He's in France at the foot of an enormous mountain and he is miserable. He has come here for snow. He's imagined it, dreamed it, dressed for it, plucked up courage for it, and there isn't any. There's just dead grass and a howling wind.

The picture is of my youngest son. It cheers me up not because I revel in others' misfortune but because of what happened before and after it was taken. A month earlier, the Met Office in London had issued a chart that sent a ripple of excitement through the defiant community of people who spend the summer yearning for snow. The chart covered the northern hemisphere from the eastern United States to the Urals. It showed two areas of deep low pressure: one over Denmark, the other over Corsica. Between them, slanting for a thousand miles from north-east to south-west, were three cold fronts looking like a diagram of invading armies in an old history text book. In the middle of their advance lay the Alps.

The chart was a forecast for November 5th. That day, right across the arc of mountains that separate Italy from the rest of Europe, it started snowing. The snow continued for three days. In the higher French and Swiss resorts a metre fell. Cars were buried as if it was February. Even lower villages woke up on the 7th to find winter had come early.

For the whole of the next week the air over the mountains stayed cold. *The Times* reported the best early ski conditions in 20 years and a lot of people went a little nuts. I know this, being one of them. I booked a weekend trip to Chamonix. I couldn't afford it but I couldn't help it either. Seize the day. Live like a lion. How much longer will the kids be around? At times like these the justifications seem so obvious.

And then the wind shifted. The low pressures wandered off the map and the temperature in some valleys rose to 20°C. The process was easy to follow and agony to witness.

A metre of snow can vanish fast, especially if it hasn't been compacted yet by machines or gravity. As warmth goes to work on them the flakes on the surface lose their crystalline structure. They turn into dull, predictable water droplets and get dragged earthwards through the snowpack, destroying more flakes as they go. A few drops following the same path create a tunnel, and soon a rivulet. When the ground below is still warm from the summer the snow melts from the bottom too, reproachfully.

By mid-November the mountains were bare again except for the summit slopes; a landscape in waiting. The skiers came anyway, trusting that their faith would be rewarded with another storm, but it wasn't. We were just slapped in the face by the wind.

There was no point prolonging the agony. We packed to leave early, but as we loaded the car an underemployed lift man strolled over. "Try Italy," he said. "Sometimes with the fohn it's better there."

The wind that had beaten us had a name. The fohn. It rhymes with burn and comes from Africa; a giant heat exchanger that carries energy off the sands of the Sahara and dumps it in Europe without the slightest concern for humanity at play.

We didn't set much store by the lift man's advice, but we took it anyway. There was a quick way to Italy, through the Mont Blanc tunnel. The tunnel bores straight through a rock barrier ten miles wide, three miles high and 30 miles long. We had not thought much about what would happen to a wind hitting this barrier broadside on, so it seemed miraculous to emerge from the tunnel's south end into a world like a shaken-up snow globe. Snow lay thick on every slope. More snow was falling steadily: big flakes but not too big. It was building up quickly and would stay soft, at least for a day. In 20 minutes we had travelled from dry autumn to deep midwinter. The sky to the south was full. Banks of cloud moved up the Val d'Aosta until they collided with the great south face of the mountain. There they stopped and piled on top of each other like airships in a holding pattern. The border police took an age with our passports but it didn't matter. When it's snowing, nothing matters.

It wasn't rocket science at work. It was our old friend, the Clausius-Clapeyron equation. The warmth of the fohn wind meant it could absorb enormous quantities of moisture as it crossed the Mediterranean. As this air moved into the mountains the gain in height cooled it down. Mont Blanc left it no-where to go but up. The moisture froze and crystallised and fell. As the crystals fell they grew into snowflakes, linking up with others to form bigger ones and sucking the air dry.

So Italy got the snow. The air that carried it there had to keep moving to make way for more and dived down into France, scouring the contours of Chamonix's deep valley like an angry ghost. But we'd left all that behind. My eight-year-old was smiling. We could not believe our luck.

It felt like winning against the odds. It was as exciting as the times you raise your eyes above the haze to see snowfields hanging cloudlike from a summer sky. It was as perfect as the first time I saw serious snow, aged 14. A glamorous aunt took me to an eerie a mile above Lake Geneva in a midwinter blizzard and we finished the journey in a horse-drawn sleigh. The idea was to learn to ski but for the first three days the snow was too thick for anything but laughing at and throwing. It was miraculous, and that is how finding snow will start to feel for those of us who live below the 60th parallel. It will take a little more luck each year. There will be more bare mountainsides, more snow that turns to rain than vice versa, more fickle fohns. And yet there are ways we could save it. There are blizzards yet to come that will be among the most intense in history, and there are snowy places still barely explored.

The obvious candidate for all-time snow valedictorian is Mount Baker in north-western Washington state. Not many places can match the four-month festival of snow that hid the sun for weeks on end and left whole houses buried until summer, on Mount Baker's slopes in 1999. To this day it claims the 1,140 cumulative inches or 95 feet that fell that winter is the world record for a season.

Do we leave it at that? Mount Baker is not especially high.

Where in the world do mountains rise higher, close to generous moisture sources, cold enough to snow? There are contenders in the Canadian Rockies and the Russian Far East. In Japan and Norway there are coastal mountains whose unobstructed exposure to the sea compensates for their modest height.

Where else?

Matthew Clarkson, a member of the Stormtrack extreme-weather-chasing community, became intrigued by this question a few years ago. He devised a model into which to plug all the data available from the US satellite-based Global Forecast System (GFS) to help identify the snowiest place on Earth.

His hunch was that Mount Fairweather and Mount St Elias, both in British Columbia, would end up duking it out for the top spot, possibly with Mount Baker, possibly just with each other. And indeed they all featured high on his list when his model spat out its results. But the winner was dramatically elsewhere. It wasn't in North America at all, or Russia or Japan or Scandinavia. It wasn't in the Caucasus or the Arctic or Antarctic. It was a mountain scarcely climbed or even visited by outsiders in the past 20 years because of civil war and the rigid determination of the locals not to be disturbed. It was higher than most peaks in North America and higher than all in Europe, although nowhere near as high as the Himalayas. It was within sight of the bath-like waters of the Caribbean and its name was Pico Cristóbal Colón.

Pico Cristóbal Colón is the highest mountain in Colombia, rising out of the jungle in the country's far north, barely half an hour in a turbo prop from the Caribbean suntrap of Aruba. It forms the northern end of the high Andes. Study it on Google

Earth and its summit cone is, indeed, as Clarkson suspected, snowy. But how snowy?

The peak is 5,787 metres high, and always snow-covered. Being equatorial it does not have four seasons. Instead it has six months of heavy precipitation from March to October and barely 20 days a year of no precipitation at all. The freezing point seldom moves far up or down from 4,900 metres. Below that it almost never snows. Above that it almost never stops. That is the verdict of climate models based on GFS data, at any rate.

The numbers are remarkable. Once I had come across Clarkson's theory I couldn't stop checking the mountain's snow forecast, which regularly predicts dumps above 5,000 metres of more than two metres *a day* even in July. That is the CF Brooks theoretical maximum. It is as much as many Alpine resorts expect nowadays in a season; an unfathomable amount of snow. It may be implausible, but the idea that the snowiest place on Earth could be a short drive from the fleshpots of Medellín and essentially unacknowledged 17 years into the 21st century was still, as they used to say in hitch-hiking circles, a trip.

In my mind I've made a pilgrimage to this place. In reality I haven't. The Kogi Indians who have inhabited the jungle around the peak for thousands of years revere it as the centre of their universe. They don't let outsiders up it. A small number of trekkers with approved guides climb each year to a set of pre-Colombian ruins known as the Ciudad Perdida (Lost City) but there is only one route. It goes nowhere near the summit and there are tales of kidnappings to stop people straying from it.

To sit with my tongue out in a two-metre-a-day blizzard on the north-east slopes of Pico Cristóbal Colón five and a half thousand metres up sounds wonderful, but being kidnapped doesn't. It would feel foolish. It would feel as foolish going there and finding that the precipitation models weren't accurate and the mountain wasn't especially snowy after all, and both these scenarios are plausible. I tell myself it's actually better to hold onto the idea of a secret South American snow stash*, undisturbed by humans and unbelievably bounteous, and not enquire too much further into the reality. Because sometimes reality is too bleak and you have to look away.

The science of global warming due to greenhouse gases is not new or controversial. The principle that atmospheric gases trap heat and make life comfortable was established in the mid-19th century by the Irish physicist John Tyndall. Without them, NASA estimates that the mean global surface temperature on planet Earth would be minus 18°C rather than plus 15°C.

Water vapour is the most abundant naturally occurring greenhouse gas. Carbon dioxide is the second most abundant. Pump more of it into the atmosphere artificially and it stands

---

* As it happens there is good evidence for such a stash, although not entirely undisturbed and not in Colombia. It is at the other end of the continent on the upper reaches of the Tyndall Glacier, on the Southern Patagonian Ice Field. Here, in late 1999, a Japanese-led expedition drilled a 46-metre ice core which on analysis indicated a staggering average annual accumulation rate of 17.8 metres of snow-water equivalent. Multiply by ten for a rough total of accumulated snow. This is many times Mount Baker's US record for a season, and is accounted for by the fact that Patagonian winds off the Pacific tend to be super-saturated even when below freezing, and at the top of the Tyndall Glacier, 1700 metres up, it seldom stops snowing.

to reason temperatures will rise. As they do, complex changes occur in ocean circulation and wind patterns, and simpler changes make themselves felt in people's lives. Winters tend to start later and end sooner. Freezing points rise. More of what would have fallen as snow falls as rain. Average snowpacks thin. Come the summer, the effect is plain to see in depleted reservoirs. Farmers and householders adjust accordingly. This is not heresy or groupthink. It's just what's happening.

The evidence is clear to anyone who prefers not to outsource information gathering and analysis to the divine. On May 9th, 2013, infrared air-sampling instruments at the Mauna Loa observatory in Hawaii, 11,000 feet up and far from anything that might distort their readings, counted more than 400 parts per million of carbon dioxide for the first time in the modern age. Polar ice cores suggest this was the highest recorded CO^2 concentration at a location that can be taken to represent the atmosphere as a whole (rather than, say, in cities, where CO^2 levels are much higher) in three million years.

A plaque at the entrance to the observatory shows a graph of CO^2 levels recorded there rising steadily from 315 parts per million in 1958 to 380 in 2005. There was nothing special about the number 400 except that it was round and roughly double the average pre-industrial level, and many people had hoped we wouldn't get there. But we did and we've gone on up. The count in 2018 is about 412.

In the meantime average snowfall in most parts of the world has gone down.

At the monastery of La Grande Chartreuse north of Grenoble, the monks do not talk to outsiders (as I discovered after driving there one spring hoping to persuade them to) but they do measure snowfall in their mountain fastness. There was 30 per cent less of it in the three decades after 1980 than in the three before. The snowpack has thinned by half and the snow lies on the ground for 30 fewer days per year.

The monastery is 945 metres above sea level. At that level snow is in retreat across Switzerland as well, but even way up at 3,000 metres and higher not one Swiss weather station has recorded an increase in average snowfall over the past 70 years.

In Bariloche, the Patagonian mountain town where Nazis went to hide after the war, it used to snow prettily every winter. Not in the past five years. In Alaska, where the 1,000-mile Iditarod dog race used to pull sleds over nothing but crisp snow from Anchorage to Nome, they tend to move the start these days to Fairbanks, 300 miles further north. In Mora, where Sweden's biggest cross-country ski race is held each year in the first week of March, they rely on artificial snow that has to be helicoptered in when the course gets too soggy for trucks.

Itemising snowlessness is painful but it feels necessary. Most of the time I prefer to clutch at straws, and the Clausius-Clapeyron equation is my favourite. This, to recap, is the one that says the amount of water vapour a parcel of air can hold rises by about 7 per cent with each degree of temperature. Over the years I have collected respectable people willing to confirm that this implies snowstorms could get heavier before they fade away.

For instance:

"In the simulations I've analysed you can get some quite big blizzards up until the year 2040."

Professor Raymond Pierrehumbert, Halley Professor of Physics, University of Oxford.

"It is indeed possible that Mount Baker, or a place in Norway or British Columbia with copious amounts of snow at mild temperatures, could experience an increase in seasonal average snowfall and therefore also the possibility of breaking the 1998–99 record."

Dr Philip Mote, Oregon Climate Change Research Center.

"As long as temperatures remain right you can actually end up getting bigger snowstorms."

Dr Kevin Trenberth, US National Center for Atmospheric Research.

I spoke to Trenberth in 2012. Three not particularly snowy years later he was no less confident about basic atmospheric physics: "Going forward," he wrote, "in mid-winter, climate change means that snowfalls will increase because the atmosphere can hold 4 per cent more moisture for every one degree Fahrenheit increase in temperature. So as long as it does not warm above freezing, the result is a greater dump of snow."

Sceptics seized on comments like these as evidence of a climate change mafia tying itself in knots to explain what was otherwise inexplicable given their world view. I seized on them much more selfishly, hoping they mean I will see enormous snowfalls in the rest of my lifetime even if my children don't. At the same time I know I'm deceiving myself. I'm wallowing in confirmation bias, tuning out what I don't want to hear and noting only what I do.

Context is everything. Moments before Kevin Trenberth uttered his longed-for words about bigger snowstorms, he reminded me that "as time goes on more of these storms will be rainstorms rather than snowstorms". Philip Mote was drawn only reluctantly into speculation about Mount Baker's season record being broken. He started by saying: "My educated hunch is that there is little basis for the narrative that blizzards will get stronger." As for Professor Pierrehumbert, here's the full quote: "In the simulations I've analysed you can get some quite big blizzards up until the year 2040, but between 2040 and 2080 it starts to get too warm to have much snow at all and it gradually sort of peters out."

It gradually sort of peters out.

Six years after he said that, I got back in touch to ask if the outlook was any sunnier, or snowier. He explained that the 2012 projection was for the American Midwest, based on work he'd done for a Chicago weather station to try to explain why the snow cover was thinning for cross-country skiing in northern Wisconsin. But no; no change to the general picture looking ahead to the rest of the century. Only higher resolution models and more confidence in their findings. After 2040, he said, snow will still form high above the prairies, "but you reach the point where the lower atmosphere is warm enough that it all melts on the way down".

Way back near the beginning of this century I went for a walk on Mammoth Mountain, California, with two toddlers (both mine). Snow was everywhere. They loved it. One of them picked up a lump of it and said: "Look what I found!" How we laughed. It was like finding hay in a haystack, but the time will come when finding snow at the latitudes where most people live will be more like spotting a rare bird.

There will be a final snowflake, just as there was a first one. It may not start its earthward journey for millions of generations, closer to the time when the sun expands to evaporate the oceans and consume the planet. Or it may come sooner. Can we push it back?

Sure we can. I say this because it beats moping and because stories sometimes come along that allow a snow junkie to hope. One of these I wrote myself. It was an interview with Professor Stephen Emmott, the scientist hired by Bill Gates to run the Microsoft Research Centre in Cambridge. He was pessimistic about humans' ability to slow their birth rate to a sustainable level, but he was puckish on climate change. He had put a team to work on artificial photosynthesis, and he had a dream of scaling it up into a forest of a trillion artificial trees. These would create energy as plants do when they photosynthesise. In the process they would absorb carbon dioxide from the air on a colossal scale.

Even I could grasp the essence of the idea. The graph at the entrance to the Mauna Loa observatory is, from a distance, a steadily rising line to represent carbon parts per million in the atmosphere. Viewed closer up it is a rising zigzag, with sharp downward zags each spring when trees come into leaf across the northern hemisphere. As they do this they take carbon from the atmosphere much faster than for the rest of the year. The holy grail of artificial photosynthesis is to have this process running full time. It could happen.

Another story to lift the spirit as autumn and spring eat away at winter came from British government carbon dioxide emissions data in March 2018. The level of emissions from fossil fuel consumption was lower than in 1890 and 38 per cent lower than in 1990, thanks largely to a steady fall in the quantity of coal consumed by power stations. To this sort of news the standard response from those willing to let snow go the way of the Lorax is that China is building coal-fired power plants rather than shutting them down, and in that context efforts made by small countries like Britain are insignificant. I disagree. All efforts matter. They can prove concepts and offer examples. And no one wants to cut air pollution and promote skiing quite as urgently as President Xi Jinping.

No one that is, apart from the eight-year-old who sat on a low stone wall at the foot of a French mountain yearning for snow. He's ten now, and catching up with his brothers on skis, and he would be grateful if we could all get stuck into the life-affirming job of saving snow. If we don't, we'll miss it when it's gone.

SNOW Q & A

❄ ***Is it true no two flakes are alike?***

Absolutely. See p.38.

❄ ***Is it true they all have six branches?***

Not quite. See p.39-42.

❄ ***How many flakes are needed for a snowperson?***

About 100 million. See p.16.

❄ ***How many flakes fall on planet Earth in an average year?***

About 315,000,000,000,000,000,000,000. Also p.16.

❄ ***Is southern-hemisphere snow different from northern?***

No, although some people think it is.

Joe Simpson, the climber and author of *Touching the Void*, wrote that "South American mountains were renowned for these spectacular snow and ice creations, where powder snow seemed to defy gravity and ridges developed into tortured unstable cornices of huge size, built up one on top of the other". Simpson became convinced that the snow in Peru's Huascarán range was uniquely light and treacherous after he nearly died there in 1987.

Xavier Delerue, one of the world's best snowboarders, claims Antarctic snow clings so well to icebergs that you can ski 70-degree slopes on it without causing an avalanche.

In fact there is no fundamental difference between southern- and northern-hemisphere snow, but a simple function of astrophysics does make the two lie differently on mountains in the spring. In the northern hemisphere, south-facing slopes catch more sun than north-facing ones and so allow grass and mud to show through earlier in the season. South of the equator it's the reverse. South-facing slopes have the best cover, and the ski lifts. For visitors from the north this can be discombobulating. You don't know how accustomed you are to finding the best snow on north-facing slopes until it isn't there.

How do you build a snow cave?

This is much easier than building an igloo, and much more likely to be useful. You do need large quantities of snow, but that shouldn't be a problem. You only need a snow cave when you have a lot of snow.

There are various methods, but the simplest I have come across is one I was taught outside the Cabane du Trient on the east flank of the Aiguille du Tour when stranded there in a late-spring blizzard many years ago.

I think of this method as the half-ring donut. All you need for it is a big snow bank or drift at least two metres high with a roughly triangular cross-section. Dig two circular holes horizontally into the snow half a metre wide and a metre apart. They should penetrate slightly more than the length of your body into the snow. Use an avalanche shovel if you have one or gloved hands if not. Ray Mears, the TV survivalist, has made the case for using a wood saw for snowcave building, but this seems an unlikely tool to pack for most winter adventures.

After digging your tunnels, close the entrance to one of them and connect them with a transverse tunnel at the far end. If time or energy is short, that's it. You already have enough shelter to survive a snowstorm. To make your cave more comfortable, smooth the ceiling with the back of a glove to prevent melting snow gathering and dripping from irregularities. Half-close the remaining entrance but sleep with your head close to it. Cold fresh air is better than carbon dioxide poisoning.

Having mentioned Mears it is only fair to summarise his method for a luxury snow cave complete with cold sink, heat lamp and twin bunks. He prefers to start by clearing away enough of the snow bank to create a vertical wall about three metres high. He then attacks this with the saw. The bank must be deep and high enough to allow the excavation of a cavity with the cross-section of a fat capital T about two metres deep. Mears does this by sawing big wedges of snow, thick end out, which can then be easily removed by hand and used Eskimo-style for an external wall with a low door to crawl through. (This requires snow with good structural integrity. Fresh powder obviously won't work; you need to be dealing with flakes that are old enough to have bonded together but not so old that they are ice.)

The horizontal portions of the T make the two bunks. Cold

air collects in the vertical portion in the middle. Warm air stays high, keeping sleepers cozy. A candle can be lit to add atmosphere and extra warmth – and safety. If it goes out, get out. Your carbon dioxide level is too high.

What is the world's most expensive ski lift?

The $130 million Skyway Monte Bianco in Italy. See p.189.

What is the world's most expensive pair of skis?

This may have been the limited-edition Lacroix model sold in 2008 for 50,000 euros a pair (see p.219), although the newer Lacroix Ultime, with bamboo core and titanium alloy skin, is not cheap at 8,500 euros without bindings or accessories. Custom-made Zai skis from Disentis in Switzerland, with cellulose acetate surface and granite core, go for up to $10,000 a pair.

Where is the world's highest snow served by a ski lift?

Chacaltaya is 1,700 metres above the Bolivian capital, La Paz, on the southern flank of the Cordillera Real. The Chacaltaya Glacier melted in 2009 but the drag lift can still open after heavy snows. It's driven by a car engine. Since all the skiing is higher than Everest base camp at 5,300 metres (18,000 feet), visitors may want to bring coca leaves to chew to boost oxygen uptake and ward off headaches.

What is the fastest anyone has travelled on skis?

On March 26th 2016, Simone Origone of Italy broke his brother's speed skiing record over a one-kilometre course in Vars, on the edge of the Parc National des Écrins in France.

Between the timing gates at the bottom of the course he clocked 254.958kph or 158.423mph. That is about 70 metres per second.

Is that faster than terminal velocity?

Not necessarily, because terminal velocity depends on drag, but it is faster than a skydiver falls having reached maximum speed in the classic belly-to-Earth position. That speed is about 196kph or 122mph.

How did he do it?

In a skin-tight scarlet latex suit and teardrop-shaped helmet on 240 centimetre skis on a 45-degree slope, hunched in a perfect tuck.

What is the fastest a woman has travelled on skis?

Not much slower. Valentina Greggio, also of Italy, reached 247.083kph or 153.530mph on the same day that Origone set the world record, in the same place, in another skin-tight scarlet suit.

What makes snow so slippery?

Mainly the very thin layer of water molecules that can remain in fluid form on the surface of a snowpack even in sub-zero temperatures. See p.69.

What is the adiabatic lapse rate and why does it matter?

The higher you go, the colder it gets. As air pressure falls, temperatures fall in line with the adiabatic lapse rate, which equates to about 5°C per vertical kilometre in moist air and 10°C in dry air. If this didn't happen there would be no snow on Kilimanjaro or any other mountain where it doesn't ordinarily snow at sea level. See p.192.

What's the difference between firn and fohn?

Firn is solid, crystalline snow that fell at least a season ago and is on its way to being ice.* A fohn wind is the warm, dry airflow that pours down the lee side of a mountain once any moisture has condensed out of it on the windward side.

What is sastrugi?

Snow blown by wind into long, wavy serrations with a hard crust. Cursed by polar explorers as some of the most tiresome snow through which to pull a sledge.

How can you predict an avalanche?

You can't, but you can reduce your risk. From the 1970s until about 2010 scientists thought it might be possible to "hear" the signs of an imminent avalanche. They thought unstable snow might emit low-frequency warnings before it started to slide. This would have been useful but in fact it doesn't happen. Tests with microphones and ultrasound machines have shown that

* Snow can easily be more than a season old and not yet firn. In 1866 the British alpinist Edward Whymper set out to study the transition from snow to ice by examining a vertical cross-section of the Stock Glacier, 11,650 feet up on the Swiss–Italian border. He dug a snow pit 22 feet deep and found that down to 13 1/2 feet and in some places to 15, the white stuff was still "decidedly and unmistakeably *snowy*; that is to say, lumps could readily be compressed between the hands".

the earliest sound you might get from an avalanche is the crack of a snowpack breaking, and by then it's on its way.

The closest thing possible to avalanche prediction comes from physically studying the snow. For this the most important tool is a snow shovel. A pair of telescopic snow probes is also useful, as is a snow saw or piece of thin rope. And you'll need a credit card.

First, dig a snow pit. Don't dig it in the middle of a slope you want to check in case it avalanches on top of you. Dig it nearby, on a similar gradient and facing towards the same point on the compass. It should be about a metre and a half wide and a metre and a half deep or down to hard ground, whichever is deeper. Make the back wall as vertical and smooth as possible. Next, check for hard layers that might cause everything above them to slide off as it did in Galtur in 1999 (see p.138-46). To do this, run the credit card through the snowpack, down from the top of the side wall and up from the bottom. In the process you may encounter soft layers too. Look out especially for granular depth hoar, often found near the bottom of the snowpack, weakening the whole structure.

Now do a compression test. Use the probes to mark out a block about 30 x 30 cm with the uphill wall of the pit the front edge of the block. Thread the rope round the outside of the probes and use a sawing motion to isolate a the block right down to the base of your pit. This can also be done with a saw. Then turn the shovel upside down and lay its blade concave side down on top of the block. Tap the blade ten times with a hand dropping only as a result of relaxing your wrist; ten more with your arm dropping from the elbow; and ten more with your arm dropping from the shoulder. Count to 30 as you do this. If there are serious structural weaknesses in the snow the block could shear and slide towards you at any point in the test. The point on the one-to-30 scale at which it shears gives you an idea of how easily the slope could go. One is easy. Thirty is hard. But of course the depth of the block is key as well: the deeper the block the greater the volume of snow that could avalanche.

Finally, do an extended column test. This is like a compression test but with a bigger block of snow, about 90 by 30 cm. The idea is to see if a break in the snowpack "propagates", or spreads. If it does, don't risk being on this slope. In fact, if the snow shears at any point in either test, it's wise to beat a retreat.

How do you survive an avalanche?

Get out of its way. Unless you have a death wish you will be wearing a switched-on avalanche transceiver (or beacon) anywhere near a potentially dangerous slope to help others try to find you if it goes. But that's for later. First, try to move. On foot, run uphill or sideways to get off moving snow. On skis, point them downhill to gain speed and then off to one side. If caught by an avalanche, hug a tree if possible. If none of these works the American Avalanche Institute recommends a backstroke-style motion to stay near the top of the snow as it carries you down the mountain. If you have airbags in your backpack, now is the time to inflate them. Studies show they can marginally reduce your chances of being buried. If not, don't panic – you'll need to keep your wits about you. As you come to a halt,bring your arms up to clear an air pocket in front of your mouth.

You may not know which way is up, but spitting can give a

clue which way is down. Reach for the surface if you can. It is virtually impossible to dig yourself out of an avalanche but any sign of you will make life much simpler for rescuers.

If completely buried, your chances of survival if rescued within 15 minutes are better than 90 per cent. After half an hour the risk of death from hypothermia and asphyxiation becomes acute. Overall, about half those buried by avalanches survive. Don't pant, and don't bother screaming. Trust your friends to find you. In snow, unless you've managed to create an air hole, no one can hear you scream.

How do you find someone in an avalanche?

This is serious stuff. Time is short, lives are at stake and a book as biased in favour of snow as this is not the place to be looking for expert advice. Dozens of climbing and backcountry skiing centres offer avalanche safety courses and time and money spent on them is never wasted. That said, here are the basics: First, look for any sign of the victim. Anything at all that narrows the field of search saves vital time. The best place to search is in a 60-degree cone starting at the last visual sign of the missing person. Switch transceivers to search mode, divide the area up between searchers and do a systematic "coarse" search moving in wide hairpins. Once a signal is found, follow it to the minimum distance reading on the transceiver. Then move left and right looking for an even lower distance. Wherever that is, pinpoint the location of the victim with avalanche probes and start digging.

Dig into the hill from below the body, shovelling snow to the side and down to save effort. When you find the victim think

A, B, C, D, E: clear the Airway, excavate the chest for ease of Breathing, restart Circulation with CPR if necessary, check for Disability (i.e. injuries, being especially careful of spinal injuries) and finally for Exposure. This person will be cold.

Why is snow so quiet?

Unlike raindrops, snowflakes make no sound when they land. As they collect – but before they start to weigh on each other and harden – they create a soft layer full of air pockets and as sound-absorbent as foam rubber. And they tend to keep humans indoors, or at least out of their cars.

How many people depend on snow for water?

About 2 billion, according to a 2015 Columbia University study. Most live near the Himalayas and the Andes, in southern Europe, Morocco and the western United States.

What is the snowiest place in Britain?

British snow lasts longest in high Scottish corries where it seldom sees the sun. The slowest to melt is usually on Braeriach in the western Cairngorms, the country's third-highest mountain. In winter the snow that collects in a cleft at the foot of Gharbh Choire Mór, a cliff on Braeriach's north face, can be 75 feet deep. Some of it usually survives the summer – snow patch experts believe it has disappeared completely only six times in the last 400 years. Unfortunately that includes 2003, 2006 and 2017, but if average temperatures fell by two degrees this patch could become a glacier. See p.179-81.

What is the snowiest place on Earth?

This is an enduring mystery with many contenders for the title, all enviable in their way and all fiercely defended by people who live near them. My verdict: the accumulation zone of the Tyndall Glacier in Patagonia. See footnote on p.251.

When did the first snowflake form?

There must have been one, alone in the cosmos if only for a nanosecond. But when? Where? It turns out not many people have given this much thought, but we know there was no snow at the time of the big bang, and there is snow now. David Christian, the Australian academic and founder of "big history", reckons the first snow in the universe was probably formed between 12 and 13 billion years ago. The first snow on Earth could have appeared soon after its creation – "and that probably means by four billion years ago". There is no hard evidence that old, but there is fossil evidence of raindrops ploughing into volcanic ash in South Africa 2.7 billion years ago, and of glaciation 200 million years before that. Most glacial ice forms from snow, so by 2.9 billion years ago we are deep into the snow age.

Where is the world's oldest snow?

Ice has been extracted from desolate central Antarctica that was formed by snow that fell 800,000 years ago. That was the world's oldest snow until 2017, when a team from Princeton University started drilling horizontally into much older "blue" ice pushed up from the depths as it flowed over mountain crests. The Princeton team's oldest cores were found in the Allan Hills area of east Antarctica. They go back 2.7 million years and one of the team has said that 30-million-year-old ice (which was snow once) could be lying undisturbed somewhere on the continent. Even if it is it would be nearly 100 times more recent than Earth's first ever glaciers, but there is no ice left from the first four of five known ice ages.

When will the last snowflake fall?

There will be one, alas, but we have no idea when. All snow will be long gone by the time the sun expands to consume the Earth about seven billion years from now, but decent snowstorms at liveable latitudes could be fading from history by the middle of this century. There is an alternative, though. There really is. See p.256.

ACKNOWLEDGEMENTS

My thanks to John Adam, Percy Bartelt, Steve Berry, Edward Brook, Christopher Burt, Iain Cameron, Marcello Campo, Ben Clatworthy, David Christian, Andrew Erath, Paul Hoffman, Jim McElwaine, Joanne Johnson, Sverre Liliequist, Charlotte Lindqvist, Colin Healey, Kenneth Libbrecht, Seth Masia, Philip Mote, Evans Parent, Marja Persson, Hans Pieren, Raymond Pierrehumbert, Vladimir Pitulko, Geoffrey Pullum, David Reay, Anton Seimon, Paul Sherridan, Norman Sleep, Sam Smoothy, Paolo Sutto, Amy Tan, Kevin Trenberth, Yusuke Uemura, Fraser Wilkin, Yige Zhang and the great, inimitable Rick Sylvester.

Thanks also to Helena Sutcliffe, Rebecca and Evie at Short Books; Bill Hamilton at AM Heath; my father for the original Allalin glissade; my aunt Lucinda for showing me my first blizzard; Andrew Dunn for enabling me to experience a few more; and Karen, Bruno, Louis and Enzo for sharing the burden of this costly addiction. Chumley's time will come.

A NOTE ON UNITS

I have switched between metric and imperial units throughout, not for scientific reasons but based on what seems appropriate. It may snow in centimetres in France, but in Colorado it just seems to snow in feet and inches. My apologies to purists from either system.

A NOTE ON SOURCES

The literature on snow is vast and mainly scientific, but any reading list should include Kenneth Libbrecht's *The Secret Life of a Snowflake*; *Snowstruck* by Jill Fredston, *Secrets of The Greatest Snow on Earth* by Dr Jim Steenburgh and Johannes Kepler's *The Six-Cornered Snowflake* (published in translation by OUP). I have also drawn on *Polar Bears*, by Ian Stirling; *Impressionists in Winter* (published to accompany the 2003 exhibition of the same name); Roland Huntford's *Two Planks and a Passion*; *Deep* by Porter Fox; *Snow in the Kingdom* by Ed Webster; *Little House in the Big Woods* by Laura Ingalls Wilder; and *Skyway Monte Bianco*, published (and built) by the Dopplemayr-Garaventa company.

There are dozens of snow forecast websites but weathertoski.co.uk is properly obsessive and looks back as well as forward. Another favourite is snow-forecast.com. The Rutgers Global Snow Lab is at climate.rutgers.edu/snowcover. Other sources that served as stepping stones in my research include Peter Robinson's 2005 paper *Ice and snow in paintings of Little Ice Age winters*; *How Much Can We Save?* by Dr Christophe Marty (2017); Douglas Powell's report on the southern Sierra blizzard of 1968, presented to the 2006 Western Snow Conference; and the British snow record started by Leo Bonacina and updated by Dave O'Hara.

Many other scientists, skiers, climbers and snow addicts have helped by answering phone calls and emails, sometimes at the dead of night. As far as possible I have made clear in the text whether quoting from secondary sources or my own interviews and correspondence.

PHOTO CREDITS

Image 1: The Downing Street mortar attack during an IRA mainland bombing campaign, London, 1991 © PA Images

Image 2: Blizzard at Cape Denison ~ Cape Denison, Antarctica, c. 1912. Frank Hurley © Mitchell Library, State Library of New South Wales

Image 3: Sahara snow, Algeria January 2017 © Karim Bouchetata/ Geoff Robinson Photography

Image 4: Wilson Alwyn Bentley, shown with his camera apparatus for photographing snowflakes © National Oceanic and Atmospheric Administration/Department of Commerce

Image 5: Frost, photograph by Wilson Bentley, 1910 © Purchase, Alfred Stieglitz Society Gifts, 2015

Image 6: Portrait of Ukichiro Nakaya, 1946 © Nakaya Ukichiro Foundation

Image 7: Plate XIX of "Studies among the Snow Crystals ..." by Wilson Bentley. From Annual Summary of the "Monthly Weather Review" for 1902 © National Oceanic and Atmospheric Administration/Department of Commerce

Image 8: Photograph of a Yeti Footprint, taken by Eric Shipton © Mercury Press & Media Ltd

Image 9: Yeti footprints on the Gangkar Punsum in Bhutan, 2016 © Steve Berry

Image 10: Oddvar Brå with broken pole in the final leg of the men's 4x10-kilometre relay at the 1982 world championships in Oslo, © Ivar Aaserud/VG

Image 11: "Drevjaskia", The oldest Norwegian wooden ski ever found © Holmenkollen Skimuseet

Image 12: *Morning after the Snow at Koishikawa in Edo (Koishikawa yuki no ashita)*, by Katsushika Hokusai ca. 1830–32 © The Howard Mansfield Collection, Purchase, Rogers Fund, 1936

Image 13: The illustration for February from *Les Très Riches Heures du duc de Berry*, manuscript illuminated by the Limburg Brothers, c. 1416; in the Musée Condé, Chantilly, Fr. © RMN

Image 14: *Census at Bethlehem*, c.1566 (oil on panel) by Pieter Brueghel the Younger (c.1564–1638) © Bridgeman Images

Image 15: T*he Fall of an Avalanche in the Grisons* by Joseph Mallord William Turner (1775–1851) © Tate, London 2014

Image 16: Claude Monet, *Snow Scene at Argenteuil*, © The National Gallery, London. Bequeathed by Simon Sainsbury, 2006.

Image 17: *Winterlandschaft*, Lucas van Valckenborch (1535–1597) © Alamy Stock Photograhy

Image 18: Snow in Trafalgar Square, London 1947 © PA Images

Image 19: Deep snow drifts in Bolton, Lancashire, 1940 © Bolton Evening News

Image 20: London Snow, author's own photograph

Image 21: Yuki-no-Otani snow canyon, © Shutterstock

Image 22: Controlled avalanche on Ajax Peak near Telluride, Colorado. Unable to trace source

Image 23: Sverre Liliequist outruns a pair of Avalanches during the Swatch Skiers Cup, 2013 © Freeride World Tour

Image 24: Rick Sylvester's leap off Mount Asgard in the 007 film, *The Spy Who Loved Me*, © Eon Productions

Image 25: Iain Cameron holding the last remaining parts of 'The Sphinx', Scotland's 300-year-old snow patch © Iain Cameron

Images 26: 'Snowmageddon' hits Washington DC, 2017 © Ryan T. McKnight / Shutterstock.com

Images 27-9 : All author's own photographs

INDEX

GILES WHITTELL is world affairs editor at *Tortoise* and former chief leader writer for *The Times*. He has written five previous books – *Bridge of Spies*, *Spitfire Women of World War II*, *Extreme Continental*, *Central Asia* and *Lambada County*. He lives with his wife, Karen Stirgwolt, and three sons in south London.